Quantum Physics and Life

Quantum Physics and Life

Ingemar Ernberg, Göran Johansson, Tomas Lindblad,
Joar Svanvik, and Göran Wendin

JENNY STANFORD
PUBLISHING

Published by

Jenny Stanford Publishing Pte. Ltd.
101 Thomson Road
#06-01, United Square
Singapore 307591

Email: editorial@jennystanford.com
Web: www.jennystanford.com

First issued in paperback 2024

British Library Cataloguing-in-Publication Data
A catalogue record for this book is available from the British Library.

Quantum Physics and Life

ISBN 978-981-4968-28-7 (Hardcover)
ISBN 978-981-5129-26-7 (Paperback)
ISBN 978-1-003-31267-3 (eBook)

Our senses make us aware of only a small part of reality. We humans move within narrow limits of classical physics. Other animal species have abilities that we lack, such as senses that detect static electric and magnetic fields.

Throughout Evolution, the biological processes that constitute Life have operated based on both classical and quantum physics. They are all part of the total physical world, a world that is explored by scientists with the help of classical and quantum physics. Our limited ability to see beyond the barriers of classical physics has hampered our progress in several areas of life sciences, leaving room for scientific hypotheses as well as philosophical and religious beliefs that are difficult to verify from experience. With the help of quantum physics, we can go many steps further and understand more of the physical world that surrounds us on Earth and extends far into the universe.

Life, Earth, and the Universe
to be explored with present and future
classical and quantum tools.
A never-ending story

Human's experience through
classical and quantum resourses
extending the human mind

Bio-organisms' extended experience
of the world

Human's limited experience
of the world

Contents

Preface

Theoretical physicists and medics formed a group, and a book was born.

The first attempts to colonize Mars have started. The quantum computer is on its way. The impact of artificial intelligence (AI) is growing. How come that quantum mechanics that was founded by Planck, Bohr, Heisenberg, and Schrödinger early in the 20th century is still not fully understood and accepted? How does it underlie our molecular functions, biochemistry, thermodynamics, perception of environment, and brain functions? Does it take part in our conscious minds? The insight is growing but the transition of explanations from classical physics is slow and resistant. Our minds do not willingly accept counter-intuitive conclusions. The outlook of the landscape looks flat but most of us accept that the Earth is a sphere. We have to add something further to our input of information *via* our senses. The landscape is not flat—the world is not governed by classical physics alone.

The year was 2012 and the Nobel laureates in physics had been presented. The medical society had invited a theoretical physicist to explain the annual prize and present the laureates to interested non-physicists and medical doctors. The physicist explained how David Wineland and Serge Haroche had shown the way quantum physics works and how it can be tested. Göran Johansson, the physicist, and Joar Svanvik, a surgeon started a discussion at the dinner table after the lecture. There were many "aha" experiences. Not classical physics alone but also quantum mechanics must be involved in biology, physiology, and medicine! Quantum chemistry is an established scientific entity. Then why are biology, physiology, and medicine still left in the realm of classical physics?

This had to be discussed further. Today's science can explain far from all in biology and medicine—maybe quantum mechanics can give further understanding. Could quantum physics, which is often counter-intuitive, give further clues to understand life sciences?

The group was extended to four researchers. Göran Wendin and Ingemar Ernberg were also affiliated. The group met regularly and took a working name, HUBIQ, Human Biology and Quantum Physics, and started a website, www.HUBIQ.se. We often met over holidays in vacation cottages in different locations in Sweden and discussed the possible roles of quantum physics in biology, evolution, medicine, and diagnostics with the intention to join non-classical physics to biology and physiology.

Brainstorming events led to a basic insight: "Even if we have difficulties in grasping quantum mechanics, Evolution, of course, has used it all the time since it started." The evolution of Life has used all available possibilities to shape enzymes to speed up processes in our bodies. It has provided us with lightning swift senses so that we may perceive the environment rapidly with minimal energy. Survival benefits—but based not on classical physics alone but also on quantum physics.

We met at an old rectory in Sörmland in Sweden for two days of lively discussions. The sun was bright in the morning when we looked out over the beautiful landscape. We noticed that in the stillness of the night, busy beavers had cut down the lovely birch tree that used to lean over the lake. The soft morning air smelled from autumn trees and decayed leaves.

Our eyesight and sense of smell, the bird's navigation, and the photosynthesis in all green plants in the world are all based on quantum physics. If our senses are using quantum mechanics, this will certainly influence the information we perceive and the reality that we experience.

In a winterly Fiskebäckskil the discussions move on in the evenings and during hiking tours by the water and along the cliffs where the artist Carl Wilhelmsson used to live. We are discussing ideas of different realities conveyed by Wojciech Zurek at Los Alamos, a reputable theoretical physicist with imaginative power, known for the concept of "quantum Darwinism." He describes the twilight zone between classical and quantum physics—the border between coherence and decoherence. One hypothesis is that there is an underlying coherent ground state of quantum mechanics— but interaction with the surroundings or measurement procedures causes decoherence. The physical events that we can see or register then land on a level of classical physics.

The dark forests on the east of Varberg are dense, as well as our discussions in the cottage. We are talking more and more about the big questions. What is Life? Are we dealing with quantum philosophy? The border between theoretical physics and philosophy is not entirely distinct. We must stick to explanations that can be verified and that are supported by logic and mathematics, avoiding thoughts that are not supported by science. We have to stay away from quantum mysticism and overly speculative ideas—the quantum world has had enough of that already.

A book is growing, slowly, meeting after meeting, while the years are passing by. The ambition from the beginning was to stay at a high scientific "text-book" level. In the end, at least for now, we chose to aim at a broader readership with a popular science version. Formulas, algorithms, and the explicit form of the Schrödinger equation were omitted, and the ideas were expressed in more popular wording.

It became evident that we needed cooperation with a scientific journalist, a communicator of science to popularize and revise the text for more accessible and easy reading. We read articles in popular science journals and found Tomas Lindblad, and he was invited to join the project. Tomas added anecdotes, simplified the wording of the text, challenged explanations, and omitted "too heavy" parts. We had found our devoted journalist.

This is the way *Quantum Physics and Life* was born. A breathtaking journey and a book that hopefully will tempt you to share the thoughts that our group of physicists and biologists/physicians have exchanged during the eight years. Welcome to the party!

Ingemar Ernberg
Göran Johansson
Tomas Lindblad
Joar Svanvik
Göran Wendin
Summer 2022

Acknowledgment

Special personal thanks to Göran Grimvall, Erik Aurell, Stephan Mangold, Andreas Larsson, Dag Cato, Enrique Solano, and Robert Langer, for careful reading and helpful criticism, as well as to families and good friends, no one mentioned and no one forgotten, for valuable views and lively discussions about and around the book.

Quantum Biology: Where It Came From

This book deals with quantum physics and its connection with life. How the smallest units of the living cell interact through the delicate forces of the subatomic world. The term *quantum biology* is often used in this context. But the concept is not a new one; it has been in use since long before there was any firm evidence of quantum effects in biological systems.

Back in the late 1920s pioneers of quantum physics like Niels Bohr and others speculated over the possibility that their new revolutionary insights into the smallest units of matter—the atom and its parts—somehow could explain the unanswered questions of biology. To many of these bold thinkers the classic laws of chemistry and physics seemed unable to fully explain the rich and complex tapestry of life in all its forms. Maybe quantum theory could shed some light on the field.

In this book, we will meet the great Erwin Schrödinger and his book *What is Life?* from 1944, where he suggested that quantum physics influenced genetics and the process of inheritance. But even before that, his colleague Pascual Jordan, a German theoretical physicist, was the first to use the term—*Quantenbiologie*—in 1932, in an attempt to unify physics and biology.

Schrödinger's ideas became an inspiration to many, while Jordan is more of an obsolete figure today. Jordan was a prominent physicist who collaborated with Werner Heisenberg among others, but he was also closely associated with the Nazi party during the war, and his tendency to mix National Socialist politics with science is an important reason why his name is less familiar today than those of his more famous contemporaries. And that may also explain why *quantum biology*, as a concept, never became a big hit in the scientific world in the post-war period. It wasn't until the Swedish theoretical chemist Per-Olov Löwdin used the term in 1963 in his description of the mechanics of spontaneous mutations in DNA that it got a slightly firmer foothold in the debate on life's mysteries.

From then on, the term *quantum biology* has slowly crept into the scientific discussion, even though many have been skeptical and questioned the real importance of "quantum" in biology.[1] This book presents some of the well-known and important instances where it does play a role and also provides a broader view of quantum physics in life. Maybe more will be uncovered further on; the history of quantum biology is still developing.

---

[1]The history of quantum biology as a scientific term is eminently described in *The Origins of Quantum Biology,* McFadden and Al-Khalili 2018, *Proc. R. Soc. A* **474**(2220), 20180674.

Chapter 1

Life and Quantum Physics

Life is really simple, but we insist on making it complicated.

—Confucius

Nothing in life is to be feared, it is only to be understood. Now is the time to understand more, so that we may fear less.

—Marie Curie

What is life? If asked that question, most people would probably tell you to come back later, much later. Or tell you right away that life is nothing that human beings should try to understand, and absolutely not manipulate. Some might argue that life is much too complicated to have emerged from Darwinian evolution—there must be intelligent design behind it. It could not have evolved—it must have been created as is.

The bottom line is: can we humans understand the nature of life to the extent that we can create living biological objects? Some of us would answer yes, most certainly—it is "just" a matter of time. And also, of course, of the very definition of life. Others would be humble and say who are we to know if science can generate the right models and tools to understand life.

But what does quantum physics have to do with life? Life is biology, and quantum physics deals with atoms and elementary particles. No connection, right?

Quantum Physics and Life
Ingemar Ernberg, Göran Johansson, Tomas Lindblad, Joar Svanvik, and Göran Wendin
Copyright © 2023 Jenny Stanford Publishing Pte. Ltd.
ISBN 978-981-4968-28-7 (Hardcover), 978-1-003-31267-3 (eBook)
www.jennystanford.com

In fact, there is, in many ways. Quantum physics rules many parts of the universe: it was essential for the start of evolution and it governs all of molecular biology and biochemistry, for example the function of DNA, proteins, and enzymes. It is fundamental to the cornerstones of life—the harvesting of energy from the sun (photosynthesis) and the utilization of energy in cells (respiratory chain in mitochondria); it provides molecular receptors in our eyes that make us register photonic quanta and see the world around us; it controls biological clocks; it provides protein ports in the cell membranes that enable the cells to exchange molecular signals; it gives us classical non-quantum humans the ability to synthesize molecules and drugs that influence our cells at the quantum level, to kill viruses or induce psychedelic effects.

The universe as we know it was born in a Big Bang 13.8 billion years ago. At the instant of creation, it might have consisted of an extremely hot soup of strings in 10 dimensions. After cooling off for a picosecond, quarks and electrons emerged, and a millisecond later the universe was converted to protons, neutrons, electrons, and photons in a more familiar four-dimensional world. It then took some 300 000 years for the universe to cool down enough so that hydrogen and helium atoms could form. Up to this point, it was all about quantum physics, but once the simplest atoms were formed, gravitation got its chance to begin pulling the atoms together.

Over the next 400 million years gravitation caused the hydrogen and helium gas to break up into dense hot regions forming the first stars and galaxies. Inside the stars, nuclear reactions burning hydrogen and helium produced all the natural chemical elements in the periodic table in a process dominated by quantum phenomena. And when those early stars became unstable and exploded as supernovae, their contents of chemical elements were ejected into space, and the stars themselves became supermassive black holes visible as quasars. In this way, later generations of stars and solar systems were born in gas clouds fueled with lots of chemical elements, and galaxies could form around central black holes.

This means that over billions of years after the Big Bang, all solar systems have been born in gas clouds containing all the chemical elements that we know on Earth. This also goes for many simple organic and inorganic compounds that are found far away out in the dust clouds in space, or in comets and meteorites. These compounds are governed by forces of quantum physics.

All the conditions and components that started the evolution on Earth must have existed all over the universe already around 5 billion years before it happened on Earth, before our Earth was in place. Life on Earth might very well have evolved from scratch starting with elementary biomolecules coming from space or forming in hot vents in volcanic regions of the deep sea. It is however not unthinkable that life on Earth was kick-started by complex life-like organic compounds already present in the gas cloud forming the solar system. One can even speculate that extra-terrestrial bacteria could have started the evolution of life on Earth.

Quantum physics describes how chemical bonds work, why hydrogen and oxygen spontaneously become water molecules, and what it is that keeps oil and water from mixing. Quantum physics explains how electrons move around an atomic nucleus, and what it is that drives transformations of matter. Quantum physics is basically behind all chemical activity.

But can the quantum processes really exist in the warm, wet and noisy environment of living cells? In fact, they can, and they do. We will describe how quantum physical processes are an essential part of the life of billions of organisms. Quantum physics makes us see the world around us, and it is perhaps behind some of the world's most advanced animal navigation systems. It may also be involved in our sense of smell. Quantum physics drives photosynthesis and makes all the world's rainforests grow. These are a few examples, and we surely don't know yet how many more there are to find. But from the Big Bang to the present-day world teeming with life, constantly growing and reproducing, quantum physical processes have played a vital and indispensable role—without them we wouldn't be here.

Chapter 2

Our World Is Just One Part of the Whole

If quantum mechanics hasn't profoundly shocked you, you haven't understood it yet.

—Niels Bohr, Nobel laureate in physics, 1922

All living things—animals, fungi, and bacteria—exist in worlds of their own. Each species has its own set of senses for survival and reproduction. We humans are just one of them, a very special mammal that experiences the world in a way that suits our special needs. Like all other living beings, we experience the physical world through waves and vibrations. Eyesight and temperature through electromagnetic waves, hearing through mechanical waves, touch and heat through vibrational sensing, and smell and taste possibly with the aid of molecular vibrations. All this information is transformed by sensory organs into signals that reach the central nervous system where it is handled and combined with internal information coming from inside our body. That is how our subjective view of the world is formed.

The world is bigger than anything we can see, hear, and feel, and most of it is hidden from us. That's because we're biological beings—the result of millions of years of evolution. Our senses form the perception of our physical world. Other species with other sets of senses experience the world differently. Our eyes can perceive electromagnetic waves in the form of light, but only in a small section

Quantum Physics and Life
Ingemar Ernberg, Göran Johansson, Tomas Lindblad, Joar Svanvik, and Göran Wendin
Copyright © 2023 Jenny Stanford Publishing Pte. Ltd.
ISBN 978-981-4968-28-7 (Hardcover), 978-1-003-31267-3 (eBook)
www.jennystanford.com

of the broad spectrum that stretches from long radio waves to ultra-short-wave gamma radiation. Outside the narrow visible range, the electromagnetic waves are not detected by our eyes. Our ears are built to register rhythmic pressure changes in the air. But only if these changes vary between 20 and 20 000 oscillations per second. If they have higher or lower frequencies, we can't hear them.

Imagine a cat is sitting in the sun on a lawn. You can see it, hear it, smell it, and if you pat it, you can also feel it. Those are our sensory impressions. But what is the objective physical reality of the cat?

Well, we can see it since there are electromagnetic waves coming from the sun, reflected by the cat, hitting our eyes, transformed by the retina into signals in optical nerves reaching our brain. We can hear it since the cat's vocal cords create waves in the medium of air, which are converted to nerve signals by the membranes and the cochlea in the ear. The smell comes from molecules picked up by the olfactory organ in our noses transforming it to chemical messages in nerves reaching the brain. We can feel it thanks to microscopic buds in the outer layer of our skin called Meissner's corpuscles that provide us with vibrational sensing.

It is all about waves and vibrations between the cat and our sensory organs. Waves and vibrations are transformed into nervous signals that travel to the brain. All these signals ensure me that there is a cat in the sun on the lawn. These sensory abilities work well for us, but other creatures have found other survival strategies. Polarization is a property of light that can give extra information beyond color and intensity. For example, about what kind of materials have reflected the light. Birds, bees, and octopuses can all see the polarization of light and use that ability to orient themselves: to find the right direction or to find a nice shiny surface of water to land on. Birds can also see ultraviolet light. That means that they have patterns and colors in their plumage that are invisible to us. Members of species where the female and the male look exactly alike to us can easily distinguish each other. The female sees more of the beauty of the courting male than the human eye can perceive.

A living body immersed in water produces a weak electric field. Sharks and rays can feel this voltage as if it were a scent, and they use that ability to locate prey. This electric sense exists even in an unborn shark in its egg. If you apply an electric voltage to electrodes near the egg, the gills of the fetal shark will stop moving, so it won't be detected by predators.

Several species of snake have special organs in the head that sense the body heat of their prey in the form of infrared radiation. Blood-sucking bats have heat detectors on their noses that allow them to find the hottest and most blood-filled body parts of their victims.

Birds, fish and sea turtles as well as many insects, seem to have a natural navigation system based on the Earth's magnetic field. There is still, however, a great deal that is difficult to explain when it comes to migrating species. For example, how can the American Monarch butterfly leave a tree in southern Canada and then migrate in stages spanning four generations until their great-grandchildren return to the very same tree their ancestors once left?

Whales shape their underwater world with sounds that travel thousands of miles—low-frequency infrasound that we cannot hear. Bats, on the other hand, map their nocturnal environment with the help of echoes from ultra-high-frequency audio signals.

Even plants that live a more sedentary life can share information with other plants. Trees and herbs emit volatile organic substances when they are damaged or attacked. A broken stalk or a chewed leaf triggers a chemical signal, that other plants sense and react to. Newly mowed grass carries the sweet scent of summer. But from a plant's perspective, it is an alarm signal from a garden massacre.

We humans do just like bats, sharks, and robins and use the resources we have been endowed with to survive and prosper in our environment. But we also have the unique ability to go beyond the limits set by our natural senses. We can measure signals that are far beyond the visible or audible. We communicate over great distances. We are getting better at understanding our world with the help of advanced instruments and calculations. And with our increased knowledge, our worldview is changing.

In the 1660s the first microscope showed that plants and animals consist of cells. A little more than a hundred years ago we learned how to see into the human living body with the help of X-rays. In 1930, astronomers discovered that our galaxy the Milky Way emitted radio waves. Radio, optical and X-ray astronomy have since completely revolutionized our perception of space. With the

Hubble Space Telescope, we can see almost all the way back to the birth of the universe, to when the first galaxies formed 400 million years after the Big Bang. As we write these lines, the James Webb space telescope is on its way to show us even more of the earliest cosmic history. In the Large Hadron Collider (LHC) in Switzerland, the theoretical Higgs boson finally appeared as a real particle. And in 2017, scientists were able to register gravitational waves for the first time from a collision between two huge black holes over a billion light-years away.

We are accumulating information far beyond the abilities of our senses, but still through waves and vibrations. Astronomy and radioastronomy study the Universe through electromagnetic waves, gravity by the effects of gravitational waves on huge laser interferometers like LIGO, and chemistry by studying the spectra of molecular vibrations. We have gone a long way past the limits of what human senses can perceive. This expansion of our senses and imagination has also allowed us also to discover quantum physics.

To describe the physiological functions in our bodies such as the circulation of the blood, the chemical messenger system of the hormones or the electric signaling of the nerves, we use explanations and models based on classical physics: pressure, flow, classical chemistry, electricity and so on. We have extended our knowledge all the way down to single molecules—molecular biology. But do these models explain it all?

The traditional model of how life works doesn't involve exotic phenomena from the book of quantum mechanics—tunneling, entanglement, and the spin of elementary particles—but Evolution seems to do just that. For instance, in our sensory organs. There seems to be a tendency toward registering the physical environment with the highest sensitivity at minimal energy consumption.

Classical physics is largely about the world that our natural senses register. It is the familiar world where apples fall from trees and the planets follow their orbits around the sun. It is the reality where the palms of our hands become warm when rubbed against one another, and where water freezes to ice when it is cold.

Quantum physics, or quantum mechanics, deals with how the smallest parts of matter interact with each other, an aspect of our environment beyond the capabilities of our senses. This model of how the world works on the smallest scale is foreign to our brains. Quantum phenomena cannot be seen. Outside the abstract models and equations of mathematics, they can only be explained by incomplete pictures and metaphors that often feel unnatural, and sometimes quite strange.

Our knowledge of health and illness is based on classical physics. Today we have penetrated far into biology and medicine with the help of biochemistry and molecular biology. We know what happens when an egg cell is fertilized and begins to divide to form an embryo. We know what paths an influenza virus takes to infect one of our lung cells and we can describe how a molecule of nicotine causes nerve cells to react in a smoking person. But many phenomena are still difficult to explain solely with the classical concepts.

Knowledge about how quantum mechanics affect our own cells and biochemistry is growing rapidly. This means that quantum mechanics' seemingly mysterious descriptions and predictions are moving into modern biological and medical research, and also into clinical medicine.

But it has taken time to realize that quantum physics really plays a role in living systems. And it has been difficult to understand how it can occur. When physicists have studied the quantum phenomena, they have mostly done so in laboratory settings at extremely low temperatures—fractions of degrees above the absolute zero—and in a total vacuum where all interfering substances and all effects from radiation, light or other electromagnetic waves are eliminated. External influences easily disrupt the fragile quantum state of matter—the quantum coherence. So how can they possibly have any effect in a plant, in a living cell, in daylight and at room temperature? "Life is far too hot and wet for quantum physics to have any influence" has been the common objection from skeptical scientists. But still, somehow the effects seem to be there.

Biologists have long been content with explaining the complex interplay of molecules in all living beings with the laws of classical physics and chemistry. But that has started to change. Quantum biology is now an established concept, and more and more of nature's processes have been found to have important quantum mechanical features. But it is still an open question whether we are talking about just a few special cases, or if we are beginning to pry open the door to life's truly deep secrets.

Today there are many bridges that connect the biological and physical sciences, but just a few generations ago it was harder to cross the divide between the two fields. Not that it was forbidden or caused by suspicion or mistrust between the scientists. Rather, because they studied such different phenomena. What could a physicist devoted to the interaction between the constituents of the atom possibly have to say about the frog's fetal development? Or about the origin of life?

Around the middle of the last century, however, the understanding of the mechanisms of life had reached such a level that it seemed clear that many of biology's most difficult questions could be answered only through the study of the molecules of the cell.

In 1944, one of the pioneers of quantum physics, the Austrian Erwin Schrödinger, published a book titled *What is Life?* It was a bold project for a physicist to embark upon, but Schrödinger's view of life's mysteries came to be influential. Schrödinger notes that life is fundamentally different from what we mostly see in nature. The classic laws of nature are based on what is called mass action. A single atom or molecule can behave more or less randomly, but many particles together follow the assumptions of probability and become predictable. The air in a room doesn't just end up in one corner. It spreads evenly throughout the volume of a room—we all know that. But a single molecule can bounce around all over the place, colliding with all the other particles in the gas. When we look at individual particles, randomness rules. For the classical law to apply, mass action is required.

You can compare this to the throwing of a die. In a single roll you can only guess if it will show a one or a six, because pure chance reigns. But if you repeat it 6 000 times, you can almost certainly say that all six sides will be on top about one thousand times each—if the die is fair. And the more times you roll it, the closer the result will follow the law of probability.

What struck Erwin Schrödinger in 1944 was that even though the processes in the living cells also depended on mass action, there were some biochemical reactions that involved only a few molecules at a time. One of those processes was the mechanism for genetic inheritance. In those days, nobody knew what the genes looked like or how they worked. But that the genetic material in a fertilized egg was somehow stored in the chromosomes of the cell nucleus seemed obvious. And this was a system that made life work with astounding predictability. The inheritance is passed on generation after generation, and it does not happen randomly. Thus, the genetic mechanisms of life did not seem to rest on the principles of classical physics, Schrödinger noted. Instead, he suggested, life was based on quantum physical events, making use of what he called "asymmetric crystals." These ideas inspired many followers, and in many ways, they would prove to be ahead of their time.

Erwin Schrödinger was a pioneer in more than one way, and he will appear several times in this book, not least as one of the creators of modern quantum physics. It did take some time, however, before quantum biology became an established concept, and it is still an open question how much of life's strange mechanisms are governed by quantum physics. To understand what Schrödinger's question "What is life?" really means, and why it is interesting to identify quantum physical effects in the great web of life, we need to find out what quantum physics really means—what it is all about.

But first, a practical example of what quantum biology can look like. We will take a brief look at a tropical phenomenon where quantum physics proves to be of great importance to a small creature.

Chapter 3

The Gecko and Life Upside Down

See that the imagination of nature is far, far greater than the imagination of man.

—Richard P. Feynman, Nobel laureate in physics, 1965

Gecko lizards are a common sight in tropical lands, where they sit quietly around the ceiling lamps, stalking the bugs of the night. They wait patiently until a prey comes within reach, and then they strike in a lightning-fast attack.

Geckos constitute a large family with more than one thousand species. Most are nocturnal with large black eyes that they moisturize with a flick of their tongue. They are unique among lizards in that they have a sound: they chirp, almost like birds. But their most striking feature is their ability to climb up smooth walls and windows, and to walk effortlessly upside down on the ceiling.

This is a trick that has fascinated throughout history. More than two thousand years ago, Aristotle expressed his surprise at the lizard's running up and down tree trunks without losing its footing. Over the years, many attempts have been made to explain how they do it. Maybe they secrete a sticky substance, or do they have some kind of suction cups on their feet? Are there microscopic wedges under their foot soles so they can secure themselves much like rock climbers? None of these explanations have proved correct. The secret of the gecko seems to be based on quantum physics.

Quantum Physics and Life
Ingemar Ernberg, Göran Johansson, Tomas Lindblad, Joar Svanvik, and Göran Wendin
Copyright © 2023 Jenny Stanford Publishing Pte. Ltd.
ISBN 978-981-4968-28-7 (Hardcover), 978-1-003-31267-3 (eBook)
www.jennystanford.com

If you look at the lizard's feet in a microscope, the intricate construction of the soles of the feet becomes visible. Under their feet, the geckos have millions of small thin hairs, only a few micrometers (thousandths of a millimeter) thick. These small hair structures are called *setae*, and it has been estimated that there are 6.5 million setae under the feet of a gecko lizard. This makes the total surface of each foot impressively large. It is the setae that are the secret of the gecko's adhesive power.

How then do thin hair-like protrusions allow the lizard to walk on walls and ceilings?

One possibility that was discussed early on was that electrostatic charges could cause the gecko's feet to adhere to different surfaces. It is a well-known fact that you can electrically charge a stick of Bakelite by rubbing it with a suitable cloth (a common experiment during physics lessons around the world). And in the winter when the air is dry indoors, you can get a huge electric shock if you slide on a floor wearing shoes with rubber soles and then touch a metal door handle. And if you rub an inflated balloon against your hair, it will stick to your head due to the electrostatic attraction that arises from the rubbing.

If geckos have electrical charges under their feet, could that create a similar attraction to the surface? No, that explanation was already dismissed in the 1930s when a German scientist named Wolf-Dietrich Dellit in his laboratory let a lizard climb up a metal plate mounted on the wall while irradiating the air around the gecko's feet with X-rays. According to the theory, the radiation should neutralize any electrical charges between the foot and the metal plate and thus cause the lizard to lose its grip. But Dellit's lizards kept climbing undisturbed, and the electrostatic explanation was scrapped. Following the negative result with the X-ray apparatus, Dellit presented the hypothesis that the lizards probably had extremely small, invisible hooks under their feet that allowed them to cling to bumps on the wall. But they do not, which is easily proven with a good microscope.

You may wonder why so much research effort has been devoted to the little geckos. The explanation is, of course, that it is a quest to imitate the fantastic climbing ability of the lizard. Imagine all the practical uses of a pair of Gecko Shoes.

And shame on those who do not persevere—the electrostatic explanation has resurfaced in modern guise, now backed by nanotechnology and quantum physics. It has been discovered that the narrow setae hairs under the feet of the gecko lizard end with extremely thin nanowires that are only five to ten nanometers (millionths of a millimeter) thick. They are thus as thin as the membrane wall of a biological cell. The gap in the contact between the nanowires and the substrate then becomes as small as the distance between the atoms in a molecule. And for such distances, we are down in the microscopic world where quantum mechanics rules.

This extremely intimate contact allows electrons to exploit the quantum mechanical phenomenon known as tunneling. Negatively charged electrons tunnel from the thin wire to the substrate. This causes the foot to be slightly positively charged while the substrate becomes electrically negative. The result is an electrostatic attraction that holds the lizard to the ceiling—the same force that can cause the hair to rise due to a statically charged balloon—created in the gecko's foot through a quantum physical phenomenon. How electrons can tunnel is a mystery we will describe in the next chapter.

It has been estimated that if all tiny hairs under the gecko's foot were activated at the same time, the attraction would support a weight of 130 kilograms—enough to lift two (lightweight) people. But if the force is strong enough to hold the entire weight of the lizard as it hangs upside down, how can the gecko loosen its grip? How come it does not just remain up there, glued to the ceiling?

It actually does loosen its foot in a few milliseconds without visible effort. The secret is a special muscle in the gecko's lower leg that allows the sole to change shape and thus its angle to the ground. This enables the gecko to easily lift its leg to take another step in proud defiance of gravity.

There are fossils of gecko lizards stuck in amber with feet similar to those of today's lizards, one hundred million years old. Evolution thus made early use of quantum physics for animal species to assert themselves and survive. The geckos use the opportunities of tunneling to create their living space where other insect eaters find it difficult to compete. In former times, up in trees and on rock walls, but more lately they have learned to use the lamps that attract insects when darkness falls.

But now, when we know so much about the feet of the gecko, where are the shoes for climbing up walls? Well, they might come, but not quite yet. There are various varieties of adhesive tape that use gecko-inspired models and methods for attaching to surfaces, for instance by using the microscopic structures called carbon nanotubes, and the development goes on. Our ability to utilize nanotechnology is just beginning to catch up with that of the lizards and Evolution.

Chapter 4

The Quantized World

Everything we call real is made of things that cannot be regarded as real.

—Niels Bohr, Nobel laureate in physics, 1922

In May 1919 there was a total solar eclipse. The sun was completely hidden by the moon for several minutes and parts of the Earth turned dark in the middle of the day. Solar eclipses have always been rare and remarkable events, but this one would have a special scientific significance. The British astronomer Arthur Eddington led an expedition to the island of Principe outside the coast of West Africa to get clear images of the solar disk as it was completely hidden behind a black moon. The photographs he took of the darkened sky also showed the stars in the vicinity of the sun. Eddington noted that the stars were not exactly where they should be according to the stellar charts; they seemed to have moved slightly out of their normal positions.

This was a new phenomenon, never previously observed, but one that was entirely in accordance with the theories of the German physicist Albert Einstein. The light from the stars had been bent by the gravitational force of the sun. This is exactly what Einstein predicted in his recently published general theory of relativity. Even a beam of light could be affected by the force of gravitation around a massive body like a star.

Quantum Physics and Life
Ingemar Ernberg, Göran Johansson, Tomas Lindblad, Joar Svanvik, and Göran Wendin
Copyright © 2023 Jenny Stanford Publishing Pte. Ltd.
ISBN 978-981-4968-28-7 (Hardcover), 978-1-003-31267-3 (eBook)
www.jennystanford.com

Eddington's results received tremendous attention. *The New York Times* wrote on its front page: "Men of Science More or Less Agog over Results of Eclipse Observations. EINSTEIN THEORY TRIUMPHS" and stated in the article under the headline that the stars are now moving around the sky according to new modern theories. In other words, it was obvious that something radical had happened to our worldview, but perhaps not exactly what it was.

What made the scientists so excited was that Einstein's spectacular ideas had now been proven real. The world was just as strange and difficult to understand as this innovative physicist from Germany claimed it to be.

Einstein presented his most important findings between the years 1905 and 1915. His theory of relativity does not merely say that gravity can bend light rays. It also states that time is relative and can run at different rates in different places. One consequence of his theories is that a pair of twins can age differently if one twin travels at a speed higher than the other. It also says that an object gets heavier the higher the speed it has. This relativistic view of the world brought with it several consequences for time and space that were difficult to understand and seemed contrary to common sense.

The theory of relativity describes a reality where speeds begin to approach the speed of light and masses are enormous, at least from our earthly everyday perspective. These are conditions that are beyond our experience. For today's astronomers, however, the theory of relativity is a necessary tool. And not just for astronomers. In the large LHC accelerator at CERN, Switzerland, the small elementary particles are driven up to such high speeds and energies that their mass becomes more than a hundred times greater than when they are at rest.

Simultaneously and in parallel with Einstein's revolutionary descriptions of time and space, quantum physics also began to take shape, especially through groundbreaking discoveries by physicists Max Planck in Germany and Niels Bohr in Denmark. Quantum physics describes the invisible microscopic world made up of atoms, molecules, and elementary particles.

The early 1900s was a time when it became increasingly clear that our everyday reality is only a small and limited part of everything that exists. The new descriptions of the world were so different that physics before Planck, Einstein, and Bohr is called classical physics, while everything based on quantum physics or relativity is called modern physics. The fact that the division still exists today, after more than one hundred years, shows that there was a radical shift in the world view—a paradigm shift.

Quantum physics was born out of attempts to understand how the smallest elements of matter interact with each other. It began with Max Planck around 1900 studying how the color and intensity of light from glowing bodies changed with temperature. It was known at this time that light is an electromagnetic wave, and that light as a wave is characterized by both wavelength and frequency, where the frequency determines the color of the light. The groundbreaking finding of Planck was that the energy of light must be proportional to its frequency, and the proportionality constant became known as the Planck constant.

However, the truly revolutionary conclusion that light waves are quantized and consist of a stream of light particles—photons—was formulated by Einstein just a few years later and published in 1905 in a paper explaining the photoelectric effect. The energy of the photon light particle is then related to the frequency of the light wave via Planck's constant. This became the first groundbreaking example of the phenomenon that matter in the microcosm can simultaneously be described as both particle and wave.

However, it took a long time for this particle-wave duality concept to be commonly accepted for waves of light, perhaps indicated by the fact that Albert Einstein did not get the Nobel Prize for his work on the photoelectric effect until 1921, over 15 years later (and never for the theory of special or general relativity).

During the 1920s, mathematical models were developed to describe how the interactions between electrons and photons worked. The results proved to be about as difficult to grasp with our everyday logic as the relativistic theories but in new ways. One of the basic but confusing discoveries revealed that the only way to understand this invisible world was that the smallest building blocks of matter sometimes appear as waves and sometimes as particles. This is one of the ideas one needs to accept to understand quantum physics: There is no simple description of the smallest parts of matter. They can be described in at least two different ways, and the two may seem to be mutually exclusive. The discovery did not resemble anything seen before. Waves and particles are two completely different phenomena in our conceptual world, but not in the quantum physical world. It is a fascinating fact that Newton's view of light as particles and Huygens' view of light as waves were finally united by quantum mechanics.

Let us now take a step back and illustrate fundamental concepts like coherence and interference in terms of classical waves. If you stand on a beach and watch the waves rolling in, you see that they come at more or less regular intervals and that they approach the beach at a speed of about a few meters per second. A wave front can pass on both sides of an obstacle, such as a rock that protrudes out of the water. When the two parts of the wave meet again on the other side of the rock, a pattern will appear if the distance between the waves—the wavelength—is comparable to the size of the rock in the water. Where two wave peaks or wave valleys meet, the wave amplitude becomes twice as large, while in the places where a wave peak meets a wave valley a zero-amplitude flat water surface occurs. Physicists say that the two parts of the wave interfere with each other and create an interference pattern.

But if the wind, or a passing boat, suddenly rips up the waves, the fine wave patterns may be obliterated. It becomes *incoherent*. The wave properties become unclear—there is no longer any clear relation between peaks and valleys. As we shall see, this can also happen to quantum waves.

A classical particle on the other hand is something small, hard and distinct. Both waves and particles have a fixed velocity. But a fundamental difference is that a particle has a position, while the wave has an extension. This means that if you throw a tennis ball (the small particle) against the same rock that previously scattered the wave, it either passes to the left or right of the rock or bounces off in some direction. The wave, on the other hand, does all these three things at the same time. This means that the pattern of interference, which occurs after a wave has passed around the rock in the water, is a characteristic of the wave, a pattern that the particle can never dream of recreating. This can be illustrated by the classic cartoon where a ski track runs on both sides of a tree.

Another difference between particles and waves is perhaps too obvious to be noticed—namely how you count them. The question "How many particles are there?" is quite normal, while the question "How many waves are there?" is much more difficult to understand. The waves form a pattern without a clear beginning or end.

   Physicists began to realize during the first decades of the 20th century that the quantum world—physics on the very smallest scales—is different from our everyday reality, and that everyday language is often insufficient to describe what is happening. In particular, they realized that it is not only the photons of light that can simultaneously act as both waves and particles. The French physicist Louis de Broglie proposed in 1924 that *all particles* can be associated with waves, with wavelengths that depend inversely on the mass and speed of the particle, and where the proportionality constant is precisely Planck's constant. This implied that the speed of a particle could be adjusted so that it behaved like a particle at high speed and as a wave at low speed.

   The Planck constant is an extremely small number, but if the mass and velocity of the particle are also small enough, the wavelength of the particle can be comparable to, for example, the size of an atom. In 1914, Niels Bohr had invented his model of the hydrogen atom where electrons orbit around a proton-nucleus in stable orbits due to quantization of the rotational motion. With de Broglie's wave picture from 1924, this quantization is quite easy to understand: the electrons are waves, and only those orbits are stable where the electron wave "bites its own tail," performing an integer number of wave oscillations on its way around the atomic nucleus. The atomic electron can therefore not understand the joke in the cartoon with the ski tracks that pass on both sides of a tree. It does it all the time with the atom. If the same electron instead greatly increases its speed, or if one considers a much heavier particle, the wavelength may then become far too small to create any interference pattern. The particle will then bounce like a tennis ball, and thus behave just like—a particle.

   There is a famous experiment where the Japanese physicist Akira Tonomura pedagogically shows that electrons, just like the photons of light, behave as wave motions. He sent individual electrons through a vacuum tube against a small metal rod, which the electrons cannot pass through but must go around. The electrons then hit a screen where they lit up like small dots. The video from the experiment shows how the electrons first appear to form a random pattern of points. But when enough electrons have passed the obstacle, a classic beautiful striped interference pattern arises, a result of interfering waves. The electrons are thus dot-shaped when they appear on the screen one by one, but obviously at the same time go both to the right and to the left of the obstacle, just as a wave does.

Another of the theoretical cornerstones of quantum mechanics is the uncertainty principle, sometimes named after its author, the German physicist Werner Heisenberg. The principle says that we cannot say exactly where a particle is located at the same time as we determine how fast it moves. And this is not due to our inability to measure, but it is a fundamental property of matter at this microscopic level. This is how the smallest particles of matter behave. One way to explain the uncertainty is that the particle is a wave, that is, it is extended in space and does not have a single well-defined position. It is only during the measurement that its position becomes determined. But when you measure the position, you lose the opportunity to measure the speed—and vice versa. An electron "is located in" a certain place only when we observe it. Before that, it exists in a more widespread state—neither here nor there, though really here *and* there. The particle is said to be delocalized.

Thus, the electron wave does not have a definite position, but its location is determined by a mathematical expression called a wave function. A physical image of an electron in a hydrogen atom shows a cloud around the atomic nucleus, a part of space where you have the greatest chance of encountering it. The wave character gives the electron an opportunity to leak into places where it should not be found according to classical physics. Classically, if you put an electron in a box, the electron will bounce against the walls like a tennis ball without being able to escape. But according to quantum physical theory, where the electron can be a wave, it has a discernable probability of sometimes being outside the box if the walls are not too thick. The electron wave reaches beyond the walls.

Translated into a more mundane parable, it is as if one would bounce a tennis ball against a wooden board, but that the ball occasionally disappears through the wall and ends up on the other side. This phenomenon is called tunneling, as if the electron utilizes an invisible tunnel through an impenetrable wall. Tunneling actually happens in the quantum world, and in addition, it is quite common and a fundamental phenomenon in the realm of life. We saw an example of it in Chapter 3—in the gecko's foot. And there are more examples to come.

The calculation of the probability of where to find a particle in the box got its mathematical expression by a man we have already encountered in the beginning of this book: the Austrian physicist Erwin Schrödinger. His description of the electron as a wave was presented in 1926 with the mathematical expression that came to be called Schrödinger's wave equation. Schrödinger based his work entirely on de Broglie's wave theory. Heisenberg, on the other hand, thought of the electrons as particles. Nevertheless, the two ways of describing the world at the atomic level soon turned out to be completely equivalent. Waves or particles? The answer was: both. Heisenberg's so-called matrix mechanics and Schrödinger's wave mechanical model were interchangeable. Heisenberg received the Nobel Prize in 1932, Schrödinger the following year.

Another central concept in quantum physics is *entanglement*. This implies that two particles can be created so that their fates are connected. For example, there may be two particles that are "born" from a collision at the atomic level—perhaps electrons or photons. If you then affect one of them, the other will also be affected at the same moment, regardless of whether the two are far apart. This seemingly unreal principle of entangled particles has been proven time and time again even though Einstein himself doubted it and claimed that the theory of entanglement proved that the concepts of quantum physics were not fully developed.

Two particles of light, photons, that travel in different directions will soon be far apart—after all, they move at the speed of light. Let us assume that we detect one of the two photons, and we find that it has the property of being polarized in direction A. Then, according to quantum physical theory, we can definitely say that the other particle is polarized in direction B. (Polarization is a property of light and other electromagnetic radiation that is used, for example, in some types of sunglasses.)

This may not sound so strange. One can compare with a pair of gloves that are lost. After some searching you might find one of them, a left-hand glove. Then you know instantly that the missing glove is a right-handed one. No matter how far away it has happened. No information needs to be passed between the gloves for us to get it. The information is already built into the system "a pair of gloves—it is a question of built-in symmetry."

Quantum mechanical entanglement is like the gloves but is far more mysterious. According to the principle of uncertainty, neither of the two particles has any definite value of the property we are interested in detecting, in this particular case the polarization. They have both possible values (A and B) at the same time. The result we get when the measurement is made is as random as when we toss a coin. It will be either A or B. But the very moment we get a result for the first particle, we know the result for the second particle.

This can never happen with two coins, but it happens every time with the particles. So, the question is: how can the second particle know the result of the first measurement? The "simple" answer is that this is the way nature works, quantum mechanically. Entanglement is based on the need to describe the state of both particles simultaneously in quantum physics. But since entanglement is a genuinely quantum mechanical correlation between the two particles, it is sadly not possible to explain their relationship in common everyday language.

Another quantum mechanical phenomenon that we will encounter is *spin*. Spin is a way of describing a property of electrons, and quantum particles in general, not found in the classical description of the world. The spin is an imagined rotational motion of the electron, like a classical top spinning around its axis. However, the quantum spin rather involves a magnetic property that determines the energy of an electron in a magnetic field. The electron spin behaves a bit like a magnet needle with north(N)–south(S) poles. When one detects an electron, only two different spin directions can be observed: the magnetic N–S needle is either oriented along the applied magnetic field or in the opposite direction. Electrons share this property with both protons and neutrons, and they are all said to have "spin ½," where the two allowed values plus or minus ½ correspond to pointing along or opposite the magnetic field.

Particles with spin ½ obey a principle named after the Austrian physicist and chemist Wolfgang Pauli. He was born in 1900 and at the age of 25, he formulated the principle that would later give him the Nobel Prize. This principle simply concludes that two spin -½ particles cannot occupy the same state. It is an extreme version of the Western movie statement "this town is not big enough for the two of us" made into a natural law. So, if one electron is occupying the innermost orbital around a nucleus, the next electron will have to settle for the next orbital, with higher energy. Without this principle, all electrons would bunch up in the lowest energy orbital close to the nucleus and our world would indeed be very different. Now, the different chemical elements of the periodic table have different number of protons and thus different number of electrons to be electrically neutral. The more the electrons, the more the different orbitals further away from the nucleus that need to be occupied. The chemical properties are to a large extent determined by the outermost electrons, that most easily interact with other atoms. For certain magic numbers of electrons, the total electron cloud has perfect spherical symmetry, which is indeed a very stable and harmonic state. These atoms are the noble gases, which interact very weakly with each other and with other atoms.

Not all particles have this intrinsic urge to be unique that characterizes electrons, protons, and neutrons. The photons we see have spin 1 and they are all perfectly fine sharing the same state with fellow photons. Radio waves, microwaves, and laser light all consist of huge amounts of identical photons, which makes it straightforward to verify the collective wave properties but much harder to see that they are also individual particles.

Quantum mechanics was initially an exploration of the fundamental properties of matter. But what was once pure basic research driven by curiosity has over time also become a prerequisite for the technological development that is constantly going on around us. Modern chemistry can be described as applied quantum physics. Computer chips with billions of transistors, digital cameras with light detectors, solar cells, lasers, GPS systems, future quantum computers and much other modern technology are based on knowledge of how the smallest parts of matter behave and function at the quantum level.

What is becoming increasingly clear is that life has also taken advantage of the possibilities of quantum physics to find the best solutions to the great challenges it entails to survive and multiply on this planet. The gecko lizard in the previous chapter is just one simple example.

Quantum physics is about to become part of modern biology and even medicine. Tunneling, delocalization and entanglement are phenomena that have helped both animals and plants for hundreds of millions of years and may have played a crucial role in the emergence of life. There are even speculative theories that claim that quantum mechanics lays the foundation for our consciousness—one of the most difficult-to-explain phenomena we know of. We will get there later in the book, but first some thoughts on how life's very emergence can have quantum mechanical connections.

Chapter 5

Evolution: About the Origin of Life

In the transformation and growth of all things, every bud and feature has its proper form. In this way, we have their gradual maturing and decay, the constant flow of transformation and change.

—Zhuang Zi, Chinese philosopher

…whilst this planet has gone cycling on according to the fixed law of gravity, from so simple a beginning endless forms most beautiful and most wonderful have been, and are being, evolved.

—Concluding sentence of Darwin's *On the Origin of Species*

Charles Darwin published his theory of evolution in the book *On the Origin of Species* in 1859. There he shows that man is part of nature and on equal terms, that we are part of the animal world and not necessarily the Crown of Creation. That was no trivial claim at the time. Fifteen years earlier, Darwin wrote in a letter that it was "like confessing a murder."

Darwin made his discovery without knowing much about cells, inheritance, genes, and even less about mutations, DNA, and the genetic code, discovered almost 100 years later. This is amazing as Darwin's evolutionary theory seems to be dealing with exactly inheritance, mutations, DNA, and the genetic code. The general opinion in the Victorian age was even further removed from

Quantum Physics and Life
Ingemar Ernberg, Göran Johansson, Tomas Lindblad, Joar Svanvik, and Göran Wendin
Copyright © 2023 Jenny Stanford Publishing Pte. Ltd.
ISBN 978-981-4968-28-7 (Hardcover), 978-1-003-31267-3 (eBook)
www.jennystanford.com

Darwin's new view of the living world. For a great number of people, even among the educated classes, the world was created by God with all the species ready-made from the start, according to the Biblical narrative. And this was just 100 years after Linnaeus had classified the living beings according to the idea of an eternal divine plan. The murder metaphor referred to this removal of God as the Great Creator and the introduction of a common ancestor. Something that probably would have rendered him a death sentence at the time of the Enlightenment. It is impossible to imagine the ingenuity and boldness of this one man who cracked the enigma of Evolution, probably once and for ever, surrounded by metaphysics and religious dogmas.

In summary, the Darwinian evolutionary theory states that (1) life on Earth has one common ancestor; (2) change and adaptation have occurred gradually, not in big steps or jumps; (3) the mechanism behind these changes depends on variation combined with selection; (4) this leads to elimination of characteristics that cause diminished fitness in relation to the environment. A fifth point deals with how new species arise and evolve separated from other species.

It took decades, however, before his theory was fully accepted among his colleague biologists and natural scientists. In 1865 Gregor Mendel, a German monk in the town of Brünn (today's Brno) in Bohemia published his now-famous paper on how the characteristics of peas were inherited according to simple and predictable patterns, with dominant and recessive traits. It didn't get much attention at the time, but it is today well known by practically every high-school student. But despite Mendel's work, the concept of genes and inheritance had to be rediscovered during the first decade of the 20th century by Thomas Morgan and his colleagues, with their painstaking and meticulous analysis of how mutations were passed on between generations of fruit flies. The complete acceptance of Darwinism as the model for evolution coincided with the 150th anniversary of Darwin's 1859 book. Then Darwin and DNA became the topic of novels, plays, operas, and arts, forming a fundamental part of our culture.

There are parts in the history of life on Earth that Darwin didn't give so much attention to in his theory of evolution, and no wonder, regarding the level of knowledge in his day. Today evolutionary theory is constantly expanding, beyond what Darwin possibly could achieve. Life's diversity is almost endless. It is easy to forget the wonder of it all when we watch a gecko climb up walls to hunt for mosquitoes, or when we look skywards to follow the flight of a crow. We have become so accustomed to the living world around us that we hardly grasp what a remarkable thing it is that a spider can make a web to catch prey, or that a colony of ants can build a hill out of pine needles. It's just natural.

On a different level, in a microscopic world that we cannot see with our naked eyes, the lizard and the crow and the spider are biochemical systems of almost gigantic complexity. Just as we are. In the cells that make up all living organisms, there is a constant chemical buzz of communication between molecules, large and small, intricately constructed, working like nano-sized machines. If you take a close look at the blood in our veins you find that its red color comes from hemoglobin. It is a protein made up of more than five thousand individual atoms, an intricate three-dimensional structure with one main task: to transport oxygen from the lungs around the body to keep the conversion of energy going, every hour, day in, day out, throughout life.

Zooming in on this molecule, you will see that it looks a bit like a four-leafed flower, with each quarter carrying an iron atom in the center. When a hemoglobin molecule prepares to pick up an oxygen molecule in the lungs, the structure holding the iron atom opens slightly to create a space where the oxygen fits. Then it closes again for the journey through the blood vessels. When the molecule receives a chemical signal, the hatches are reopened to release the oxygen molecule at the destination. The binding of the oxygen molecule (O_2) to iron (Fe) is pure quantum physics—electron transfer from iron to oxygen causes the iron to move and pull at a molecular lever that causes the surrounding structure to close.

Hemoglobin consists of one thousand amino acids joined together like biochemical pieces of Lego. To make this sequence of pick-up and delivery of oxygen happen, the protein must be constructed in exactly the right way. In our DNA we have a few genes that hold the genetic code for the construction of this oxygen-transporting machine. The code is the recipe for the protein. It determines which amino acids to use, and in which order they should be lined up. The assembled chain of these amino acids then folds itself into the four-part flower-like structure that is hemoglobin: a symmetrical shape with four iron atoms in strategic locations, tailor-made for its mission.

Millions of these precision-produced molecules are created every day in the bone marrow of all vertebrates. And this is just one of many thousands of similar events that keep us alive, whether we're fish, lizards or mammals. All the time, every day. Living systems can thus create highly functional complexity, after millions of years of natural selection. The examples are endless. And it took the theory of evolution to make us begin to understand and make sense of these wonders of nature.

Darwin's story begins with the first cell, the common ancestor. He proposes that life on Earth originates from one common ancestral species. So, he did not completely dethrone God as the creator of first life. And he was wise in not trying to explain the origin of the first cell, because that is the part of the evolutionary history, which we still poorly understand. But once the first cell was in place, the rules of the evolutionary game as we know them could act more or less according to his description. All subsequent species on Earth emerged by this evolutionary process.

Put simply, evolution can be described as a logical consequence of two of life's characteristics. Firstly, individuals vary. Some are taller than average, others become redhaired, still others get a thicker layer of fat, or vary in other ways—there is always random variation in traits within a single species. Secondly, far from all the individuals born survive to reproduce. Those with characteristics that are well adapted to their living environment succeed. They pass on their inheritance to the next generation, and it is their genes that become dominant in the offspring and in the species. And if the environment changes, other traits may benefit from the new conditions.

These two principles, with the passage of time and the changes of the earthly environment, have created life's great diversity with its incredible variety of life forms and behaviors. Evolution has produced slime molds and dinosaurs, bats and mushrooms, palm trees and orangutans, not to mention bacteria, from a first simple origin around 4 billion years ago.

But how did the first living cell come to life? Darwin does not give the recipe for that. There is no way but to assume a *prebiotic evolution*, starting with the evolution of matter. During the prebiotic evolution, the quantum physical properties of the atomic world played a fundamental role. Somehow the material world has generated the right simple atoms and molecules that opened the door for evolution, as well as the right environments for life to emerge in the universe. This phase has been followed by a *chemical evolution* resulting in the transition from inorganic matter to organic substances, primitive forms of the first molecules based on carbon, hydrogen, and oxygen allowing life to start. Organic "dead" matter somehow became alive, but it is hard to say when life really started—the definition of life is a hard problem in itself.

During the prebiotic evolution, the environment of Earth was dramatically different from today. The planet rotated twice as fast as today and was irradiated every day with high doses of ultraviolet light. The Earth was covered by oceans with huge tidal waves more than ten meters high. Volcanos spewed lava and gases even under water, and meteorites rained down from space through a hot and oxygen-poor atmosphere. It was somewhere in this apparently hostile environment that life emerged.

It is a huge step from a simple accumulation of minerals and some organic substances dissolved in water to the emergence of a living cell. We know very little about how it might have happened. A cell works with so many different types of chemical processes and functions that it looks like a patchwork of mechanisms that have been pieced together from different environments on Earth during the first few 100 million years of precellular evolution.

So, where do the cells that make up all organisms come from? Water is a prerequisite for life on Earth. Our cells are two-thirds water, and it is the unique properties of water that make life possible—at least the form of life we know. One option is that life arose on the sea floors. Alternatively, it has been suggested that life's first molecules were formed out in space—even long before the Earth was formed—and then continued to develop in the oceans.

The chain of events could very well have begun with the formation of simple building blocks for life's molecules on grains of dust out in space, laying the groundwork for increasingly complicated chemical events as they subsequently landed on Earth. Amino acids have in fact been found in meteors and comets. But it is hard to imagine that molecules as complex as DNA or proteins could arise spontaneously in nature. Surely, they must be the result of hundreds of millions of years of evolution. But most biomolecules used by nature cannot be assembled spontaneously, not even given the lifetime of the universe. Fortunately, evolution is not totally random—it is a cumulative process, guided by directed chemical evolution, building on successive existing foundations.

Obviously, to keep the engine of life running you need ways to capture and store energy. And for evolution to work, a mechanism for reproduction is required—the ability to self-copy. We don't know for sure how these two mechanisms came about, but we know it has happened, and there are interesting ideas and models.

One way to attack the issue is to assume that you first need energy to harvest and build carbon-based organic molecules. This might have happened in the primeval oceans. Deep beneath the waves are openings in the seabed, hydrothermal vents, and seafloor volcanos, where hot mineral-rich lava and water flows up from the Earth's interior. For hundreds of thousands of years, the minerals in the hot water have formed large chimney-like formations with very thin walls of minerals, serpentinite. In the middle of the Atlantic Ocean stands one of these formations of alkaline hydrothermal vents, the Lost City, where greyish towers resemble the ruins of ancient, abandoned castles.

The serpentinite walls of the towers are perforated with channels where the alkaline fluids from the Earth's interior bubbles up. Outside the thin walls is the slightly acidic seawater. Thus, an acid, the seawater, and a base, the mineral fluids, create a chemical "tension" between the two sides, a bit like the voltage between the poles of a battery. Through the transport of atoms, ions, and quantum tunneling of electrons, simple organic molecules could arise and be enriched in these channels, like in a kind of molecular factory. Metals from the mineral walls of the towers could help drive the process as a kind of chemical assistants, what is known as catalysts. In this model, quantum physics in the form of tunneling of electrons is a necessary part of the first prebiotic development of life. Many proteins in our cells contain active sites with a metal atom—typically iron, nickel, or magnesium. Those metal atoms might be useful remnants of the ancient chemical factories. The hemoglobin in our blood, that uses iron to bind oxygen is but one example.

Once autocatalytic chemical processes have started, they become irreversible. There is no turning back unless someone pulls the plug and stops the evolution by cutting off the energy input of heat from the Earth's interior. The chemical processes are self-copying and, ultimately, self-organizing, that is, they repeatedly rebuild themselves, like the cell divisions that will come much later in Evolution. The system has become a network of self-propagating chemical reactions, with emergent properties, producing new and increasingly complex building blocks—like the zoo of living creatures on Earth. In this plausible but still speculative description of the prebiotic evolution, both quantum physics and theories of complex systems (network theory) are necessary for the transition steps between the early evolution of matter and the construction of a cell. Neither of these concepts—the quantum physical aspects of prebiotic evolution, nor the science of complex systems, were available to Darwin.

With the arrival of the cell, life could be organized and compartmentalized to keep chemical reactions contained and keep energy from just dissipating. How then did the cell arise? If you pour oil into water, it forms small droplets because oil and water don't mix. But if you add a little detergent, the oil "dissolves" and the drops disappear. The detergent contains substances—surfactants—that hook on to the oil's fatty chains and clump them into small invisible balls of fat that dissolve in the water. This is a consequence of how the molecules of the water and the oil are constructed. And it is also a model for how our cell membranes are constructed.

Such fat globules, or micelles, can coalesce to form elongated tubes and sheets that eventually become hollow spheres like an empty biological cell. Such a sphere—a vesicle—consists of a cell membrane-like lipid layer and it contains water, but no biological components. Vesicles are formed in nature when molecular aggregates need to penetrate a living cell. The cell membrane can then "yield" and form a vesicle that encapsulates the molecule and transports it to the appropriate recipient. These simple "cells" are thus formed by themselves under the right conditions, by self-organization.

Somewhere, and in some way, such precursors to cells can find organic molecules and encapsulate them in the cell membrane or in the interior of the cell. That this actually happens has been confirmed in several practical experiments. During a long time and under the influence of the environment, these molecules could further develop the processes that form the basis of life.

Darwin's theory of evolution has stood up extremely well to time. It has only been strengthened by the discovery of DNA, the rules of inheritance and the genetic code. Also, modern molecular biology, population genetics and viral evolution are servants to Darwinian evolution. To some extent, it can even be applied to the evolution of dead matter, of the universe and the prebiotic evolution.

Nevertheless, evolutionary theory is not a finished and closed topic. It keeps developing in many ways and forms as the foundation for the modern sciences of life. Today much attention is given to behaviors in cells that lie beyond the control of the genes, like the self-organizing principles of the immense molecular network inside a single cell—truly complex systems. Epigenetics is another field where we need to look further than classical genetic mechanisms. It was the evolutionary biologist Theodosius Dobzhansky who stated in 1973 that "nothing in biology makes sense except in the light of evolution." Few would disagree.

Chapter 6

From the Big Bang to Black Holes

Space isn't remote at all. It's only an hour's drive if your car could go straight upwards.

—Fred Hoyle

We live in a quantum world. All matter is ruled by the laws of quantum physics on the microlevel. And in every one of the living cells in our bodies, there is a constant busy activity where quantized particles interact in a myriad of ways to keep us alive. And the same goes for all other inhabitants of this planet.

The shape of the whole universe was once upon a time—billions of years ago—born out of a "quantum soup" that seethed and fluttered in a way that would lead to the formation of stars, galaxies, and planets. Without the quantum effects at the beginning of time, the universe would be a considerably more boring place. Today we know a whole lot more about how the cosmos may have started, and we can see what is going on millions of lightyears from us. Our expanded senses have truly become far-reaching.

When the space shuttle Discovery took off from NASA's Kennedy Space Flight Center at Cape Canaveral on the east coast of Florida in April 1990, it carried an instrument that was designed to revolutionize astronomy—the Hubble Space Telescope. Discovery's mission was to place this optical telescope way above the atmosphere, where it could look at the universe with clear eyes.

Quantum Physics and Life
Ingemar Ernberg, Göran Johansson, Tomas Lindblad, Joar Svanvik, and Göran Wendin
Copyright © 2023 Jenny Stanford Publishing Pte. Ltd.
ISBN 978-981-4968-28-7 (Hardcover), 978-1-003-31267-3 (eBook)
www.jennystanford.com

Optical telescopes had served science since Galilei turned his invention on Jupiter and discovered four bright moons orbiting the giant gas planet. From then on, the instruments have been refined, improved, and enlarged, seeing further and further into the surrounding space.

The biggest Earth-based observatories today are placed in the arid high plains of Chile where the skies are clear. But even in that thin air, turbulence finally sets limits to the resolution. Twinkling stars may be romantic, but not very amusing to astronomers. The noise in the air blurred the images and made it impossible to see really far into the universe. The solution was to place a telescope in space, in vacuum, where only basic physics finally limited the optical resolution. Moreover, out there in space, telescopes can see a broader spectrum of light: the far-infrared and extreme-ultraviolet segments that are absorbed by the Earth's atmosphere. The Hubble telescope would let us see further than any device before it. See things that just had been theoretical constructs, invented mainly during the first decades of the 20th century.

In 1915, Albert Einstein published his work on the general theory of relativity and gravity. There Einstein explained his revolutionary way of treating the concept of gravitation, not as a force, as in classical theory, but because bodies with mass changed the very fabric of the universe, *space-time*. Gravitation did not pull a falling rock to Earth, as Newton would have it, that is, it's not like a magnet that attracts iron. The rock rushes toward the ground simply because it follows the curvature in space-time created by the mass of the Earth, was Einstein's description. And with this view of the world followed that time, space and gravitation were closely related phenomena.

Einstein himself could verify his calculations when he showed why the planet Mercury deviated from the path that was predicted by the classical laws of motion. General relativity could explain the shift in gravitation from the sun. And with Eddington's reports in 1919 of how the stars changed places in the sky, as predicted by Einstein (Chapter 4), the theory gained the world's attention. But what did it all mean? Those few who understood the meaning of Einstein's complex mathematical equations were presented with a challenge that would lead to several dramatic theoretical discoveries.

One of the first mathematicians to embrace the new ideas was a 43-year-old lieutenant of the German Imperial army. His name was Karl Schwarzschild. He was a soldier in the Great War, and his assignment was to use his mathematical skills to calculate the trajectories of the grenades fired on the Russian trenches. But in his free time Schwarzschild sat down with the equations that expressed the relations between time, space and gravity according to Einstein's newly published paper.

One extraordinary result of Schwarzschild's work was that if the size of a star would—in theory—shrink to a certain diameter, the numbers wouldn't really make sense. The curvature of space-time would become infinite, and nothing could leave the place, not even a ray of light. This infinite gravitation would also make time come to a standstill around the star.

Our sun would end up in such a theoretical state if it somehow was compressed to a sphere with a three kilometers radius. Planet Earth would create such an infinite bending of space-time if it became a tiny ball of about one centimeter in size. In 1916 this result was more of an exotic curiosity. Today this is what is known as a black hole

When a big star burns out and finally collapses due to gravitation, it produces what is now called a *singularity* in the gravitational field, producing a region in space from which not even light can escape—bounded by the so-called event horizon defining the radius of the black hole. Later on, in 1919 when Arthur Eddington—as we saw in Chapter 4—could show how massive bodies like stars could actually deflect a ray of light, the theory became more and more accepted. That's when Einstein became world-famous.

However, when the Swedish Nobel committee decided to award Einstein the prize in 1922, he got it for his explanation of the photoelectric effect, published in 1905, and not for the theory of general relativity. There was still in the 1920s an ongoing debate on the nature of time. The ideas of relativity were by many considered too radical to accept. The chairman of the prize committee, the physicist Svante Arrhenius, had to mention it in his presentation:

There is probably no physicist living today whose name has become so widely known as that of Albert Einstein. Most discussion centers on his theory of relativity. (...) It will be no secret that the famous philosopher Bergson in Paris has challenged this theory, while other philosophers have acclaimed it wholeheartedly.

But Arrhenius also added:

> The theory in question also has astrophysical implications which are
> being rigorously examined at the present time.

Now we know for sure that there were "astrophysical implications." And these have been examined rigorously. A few years after the ceremony, the Belgian priest and astronomer Georges Lemaître predicted that the universe wasn't a static structure but an expanding vastness. And that led to the idea that it must be possible to find a beginning, by tracing the expansion backward, to the explosion of an initial singularity.

In 1929 the American astronomer Edwin Hubble could confirm this expansion after careful measurements of the light from distant stars in faraway galaxies. Hubble saw that the light was shifted toward the red end of the spectrum, a Doppler effect—like the lowering of the frequency of the sound when a train passes the bell at the road-crossing; or an ambulance passing by you in the street. A lower frequency corresponds to a longer wavelength, and that the object you are measuring is moving away from you. Hubble determined the rate of expansion of the universe, and counting backward he concluded that the original singularity, the birth of the whole universe, now familiarly known as the Big Bang, occurred 13.8 billion years ago.

Edwin Hubble is generally considered as one of the great astronomers in history, and it was natural to name the (so far) best telescope in history after him. But the Hubble telescope had a difficult start. The fabrication of the 2.4 m mirror of glass left a small error in the shape of the mirror, and once in space it turned out that Hubble was short-sighted and needed "spectacles." Fortunately, the Hubble mission was planned for a number of shuttle launches where astronauts could repair and upgrade the instruments onboard the space observatory. Even without "spectacles," Hubble was no failure though: it allowed a deeper view of the universe. But, to fulfil its mission, it had to be fixed. In the following years, one designed two additional mirrors that compensated for the error, and in 1993, two astronauts managed to put the spectacles on Hubble's nose.

The rest is history—Hubble has been extremely successful, it has been upgraded a number of times, and it is now expected to serve through the 2030's sailing in orbit 600 kilometers above the Earth. Together with other Earth-based and satellite observatories covering a wide range of the electromagnetic spectrum, from radio waves to extremely hard X-rays, Hubble has revolutionized our view of the universe.

Seeing the edge of the universe means looking into the Big Bang itself. Hubble has looked at the farthest galaxy ever seen, 13.4 billion light-years away. Today we can see it as it was 13.4 billion years ago, only 400 million years after the Big Bang. It is therefore one of the first galaxies ever formed in the universe. And at the moment of writing, we are expecting the first images from the James Webb Space Telescope, an instrument that is built to see even further back in time.

But what was before the Big Bang? Nobody knows, but there are many ideas. Some believe in the Big Bounce—the contraction of an earlier universe to a singularity and then a Big Bang.

In principle, the gravitation caused by all the matter in the universe could make the expansion slow down and finally make it come to a stop. And then everything would start to contract back to a singularity, a super-giant black hole, in a reversed version of the universal history. The present observations however suggest that the expanding universe seems not to slow down, but rather to expand at an accelerated rate. As if some internal force creates an outward pressure. This is evidence for Dark Energy, which together with dark matter presents the major riddle in our understanding of how the universe works, and what will be its future.

Dark Energy represents the "cosmological constant" in Einstein's theory of general relativity, inserting it into his equations in order for the universe not to expand—the common view at the time. Einstein later referred to it as his biggest blunder when it turned out that the universe is in fact expanding. However, now the cosmological constant is used to describe that the expansion is in fact accelerating under the action of the mystical Dark Energy.

Dark Matter—not to be confused with Dark Energy—provides the "missing mass" and missing gravitational force needed to explain for example the structure and motion of many galaxies. Current cosmological models assume that 85% of the total mass of the universe is in the form of dark matter but it remains invisible because it does not interact with electromagnetic radiation, or with the ordinary zoo of elementary particles. At least the interaction is so weak that no experiments have been able to show the presence of dark matter. It can only be detected indirectly, for instance by observing how these unknown invisible particles affect gravitation and bend the light from distant stars.

The fact that our universe started with an explosion of a singularity is problematic from a quantum point of view. Quantum mechanics "hates" singularities—there must be some internal structure under the rules of Heisenberg's uncertainty relation. A point singularity forces velocity (and energy) to infinite, and infinities are not "physical." Big Bang, black holes, and gravity necessarily must be unified with quantum physics—*quantum gravity* raises the issue of what are the smallest objects that can exist.

Already in 1899, Max Planck constructed an expression with the dimension of length based on a combination of natural constants from classical and quantum mechanics: the speed of light (c), the gravitational constant (G), and Planck's constant h. The value of the Planck length is extremely small, about 1.6×10^{-35} meters. In comparison, the radius of the hydrogen atom is the radius of the orbit of the electron circulating around the proton and is about 50 picometers (0.05 nanometers). The radius of the proton at the center is about 50 000 times smaller, about 0.9 femtometers. And the proton is made up of three quarks with radii 20 times smaller than that of the proton, about 0.04 femtometers.

Now, these quarks are nevertheless huge objects in comparison with what scientists consider to be the smallest things that can exist, or rather could have existed. To describe the initial point-like universe one must unify quantum physics, the special theory of relativity, and the general theory of gravitation.

The Planck length is considered to be the smallest possible size of a classical black hole described by Einstein's general theory of relativity and space-time. At even smaller distances, the space-time fabric is expected to show quantum fluctuations—quantum foam. The concept of quantum space-time foam was introduced by the American physicist John Wheeler in 1955 and describes limits on the accuracy with which distances and time intervals can be measured in terms of Heisenberg's uncertainty relation. At the Planck scale, objects cannot be smaller than 10^{-35} meters and time not shorter than 10^{-44} seconds.

These scales give us the theoretical size of the initial Big Bang singularity. Its size would have been 10^{-35} meters, at the immensely high temperature of 10^{32} degrees Kelvin and lasted for only 10^{-44} seconds. Fundamental forces, gravity and space-time did not yet exist. Everything was a more or less homogenous but fluctuating quantum soup—the quantum foam.

After about 10^{-36} seconds came the "inflation," driven by a hypothetical inflation field, forcing the volume of the universe to expand exponentially fast. It expanded by a factor of 10^{26} in 10^{-32} seconds. It grew more than one trillion trillion times in size in a timespan shorter than a quadrillionth of a quadrillionth of a second. Inflation indeed. After that the expansion slowed down, and the fundamental forces of gravitation and the electro-weak force appeared.

After a picosecond, the universe had finally cooled down to a mere trillion degrees and the first elementary particles began to appear—the quarks—in the form of a liquid soup of quarks, a so-called quark-gluon plasma. The LHC accelerator at CERN and the RHIC at Brookhaven have recently been able to produce and investigate the conditions at such stages of the Big Bang by accelerating and smashing together heavy ions.

The quark-gluon plasma period lasted for about a microsecond. During the following ten seconds, the universe expanded further and cooled down to a million degrees, allowing quarks to form protons and neutrons, and electrons to appear.

During the following ten minutes, the universe cooled down to ten million degrees, and protons and neutrons came together and formed the nuclei of deuterium helium and lithium, the first elements. Then, after 100 000 years, atoms and molecules began to form, and after about 300 million years, the first galaxies.

In an explosion—be it a firecracker or the young universe—all particles move away from each other, like the stretching of a rubber sheet or a balloon that's being filled with gas—the outer parts move a larger distance than the inner parts in the same time interval. As a result, in an explosion the speed of the particles is proportional to the distance from the explosion center. And all particles recede from each other, like the particles on the surface of the balloon.

But in a classical scenario, all the particles from the Big Bang would fly away from each other with the same speed, in total symmetry, and eventually be distributed evenly through space. The universe would be a place without features, without any interesting objects or any life.

Our complex and beautiful universe exists thanks to the inherent laws of quantum physics. According to the principle of uncertainty nothing is ever completely at rest. Even the vacuum of "empty space" is never empty. The inflated expansion of the Big Bang could never be totally symmetric. There would always be the small quantum fluctuations of space—the quantum foam. And these infinitesimal bubbles in the emptiness could grow with the fast expansion and make the differences that led to the aggregation of matter that would become stars and galaxies.

Therefore, quantum physics is crucial for the formation of a universe as we know it. Features like stars, galaxies, and dark matter. Around these stars are the planets, billions of them, as far as we can conjecture today. Planets that might harbor different forms of life. At least one of them does. And it would seem to challenge the laws of probability if there weren't at least a few more.

Chapter 7

As Time Goes By: The Arrow of Time

Neither can the wave that has passed by be recalled, nor the hour which has passed return again.

—Ovid

To describe the evolution of life on Earth, or the history of everything, we may draw a horizontal line on a piece of paper and mark down a number of events, from the Big Bang up to now. We then add tick marks for the appearance of the first cell, the beginning of photosynthesis, the birth of the first dinosaurs, the evolution of humans and so on. Finally, we add an arrow pointing to the right—a *timeline with an Arrow of Time*. We will realize that we are, however, only marking down a sequence of events. If we did not have any kind of clock, it would not be a timeline—it would only be *a line of events as they happen*.

Time is not an entity in itself, but it is something we measure. And we measure it by the counting of a sequence of events. This is how Albert Einstein put it in his paper on special relativity in 1905:

If, for instance, I say: 'That train arrives here at 7 o'clock', I mean something like this: 'The pointing of the small hand of my watch to 7 and the arrival of the train are simultaneous events.' So, time itself is defined by simultaneous events. But, crucially, this means that anything that can affect the simultaneity of events can therefore affect time itself.

Quantum Physics and Life
Ingemar Ernberg, Göran Johansson, Tomas Lindblad, Joar Svanvik, and Göran Wendin
Copyright © 2023 Jenny Stanford Publishing Pte. Ltd.
ISBN 978-981-4968-28-7 (Hardcover), 978-1-003-31267-3 (eBook)
www.jennystanford.com

But even if time doesn't exist as an entity in itself, the measuring of it matters and ultimately defines it. Since the birth of the Earth, the sun has been the primary clock. A clock making use of quantum effects. It has radiated quantum particles—photons—in 24-hour periods, and those photons have produced excited atoms and molecules and influenced chemical reactions on a daily timescale.

Molecules excited by this light may become extremely reactive or change shape. This will influence biochemical and biological reactions and eventually lead to the creation of biological sensors that can make organisms distinguish between day and night. And, also to orient themselves or seek targets for some interesting interactions and encounters. Eyes are examples of such high-level advanced sensors. At a more fundamental level, the oscillating influx of photons eventually led to the creation of biological clocks: biochemical reactions oscillating with 24-hour periods. Biological clocks involve quantum physics at several different levels.

Once some animals—humans—began to reflect on the short and long periods of life, they began to register the sequence of events in relation to the rising and setting of the sun, the motion of the moon and the stars, and the passing of years. Hourglasses were used to keep track of elapsed time, and the flow of water was used to drive mechanical devices that rotated at a useful slow speed.

The first all-mechanical clocks were invented in Europe toward the end of the 13th century. They were based on spring-loaded balance wheels that transferred the motion of the oscillating wheel to a stick with teeth that eventually transferred the motion to the hands of the clock. This is still the principle for small all-mechanical wristwatches.

It wasn't until the 1850's that we learned to measure tenths of seconds; a technical feat that opened new horizons in several scientific fields. Not least in neurology; now it was possible to see how fast nerve impulses traveled through the body—the speed of thought. A tenth of a second is hardly noticeable to us in our daily lives, but it is a long time in other contexts. The world's fastest sprinters compare their results within *hundredths* of seconds. A ray of light, on the other hand, travels 30 000 kilometers in a tenth of a second—equal to three-quarters of a trip around the equator.

Nevertheless, science today operates with an ever-growing accuracy, and with shorter and shorter units of time. And to accomplish that, physicists rely on the quantized phenomena within and between the smallest building blocks of matter to measure time on scales that are unimaginably small. Quartz is one of the most common types of solid material on Earth. It is simply crystalline silicon dioxide—SiO_2—that was formed from melted rock a long time ago. One interesting property of quartz is that if you bend a quartz crystal, electrical charges appear on the surfaces—it is *piezoelectric*. And, conversely, if you apply a voltage to the crystal, it bends. So, if you put a quartz crystal between metal plates and set it in vibration, like a tuning fork, you will find an oscillating voltage across the plates. A precisely manufactured crystal can act as the pendulum of an electric clock, and the oscillations can be tuned to a frequency of one per second—and push the hand on the clock face one tick forward every second.

When tiny transistors appeared in the 1960s, such clocks could be made really small, and modern integrated electronics revolutionized the field—that's when the first miniaturized quartz crystal clocks and watches became high fashion and eventually the standard form of timepiece. But the true revolution—the atomic clock—came when electronic transitions between atomic energy levels were used to control the frequency of the quartz oscillator.

The first national atomic clock standard was established at the National Physics Laboratory (NPL) in England. It was based on electronic transitions between hyperfine levels in a beam of caesium atoms. The two levels correspond to the nucleus and electron having parallel or antiparallel spins. We talked a bit about spin and entanglement in Chapter 4. Here, the caesium atom has a lone electron circulating outside a compact cloud of 54 other electrons, appearing like a very big hydrogen atom. The total spin of the caesium nucleus made up of protons and neutrons creates a magnetic field that the spin of the single outer electron must relate to—it must be parallel or antiparallel to the nuclear spin. This results in two atomic hyperfine levels with different energies, and the external microwave photons can create transitions between those levels.

The basic principle of atomic clocks is all about determining the precise oscillation of a kind of pendulum. But to measure time intervals one must be able to count the number of swings of the pendulum between two events—two oscillations. One needs extremely fast "stopwatches" that work on scales from picoseconds to femtoseconds to attoseconds, to be able to observe the vibrations of molecules, see into chemical reactions, and freeze the motion of electrons in an atom. Such "stopwatches" can be based on trains of very short pulses of laser light or electron bunches and be used in the study of the dynamics of biological systems and life processes. For instance, looking at coherent phenomena in photosynthesis.

But can a clock tick at different rates depending on outer circumstances? The answer is yes. This is the consequence of Albert Einstein's theory of special relativity, often illustrated through the "twin paradox" where a twin leaves the Earth and travels in space with a speed approaching that of light. Coming back, he finds himself much younger than his twin brother who stayed on Earth.

This is actually a phenomenon that occurs every day. Astronauts on space stations do it all the time, but they only get younger compared to people staying on Earth by nano- or milliseconds, so nobody really notices. Even people on continental flights do so, by nanoseconds. The real paradox is that time travel is not symmetric— we are all traveling through time, in the direction of the arrow. But we are not traveling at the same speed. The faster you go through space, in relation to someone else, the slower your clock ticks in comparison. The twin that left the Earth was staying younger than his stationary brother by traveling faster in space but more slowly in time. His clock ticked at a slower rate.

This can be a bit tricky to digest. It is not a thing we experience in our everyday life. One needs to remember that time is but an ordered sequence of events, to be observed and compared to the events provided by the tick tacks of a clock. This act of *observation* is key to understanding that time will be relative.

In everyday life observation is instantaneous, and two people observing the same event can both state that it happens at the same time. If the observers are very far apart, however, they will record the event at different times because the observation is dependent on the timing of the light reaching them. And if they are moving with respect to each other with velocities approaching the speed of light, the concept of events being simultaneous will become very relative.

As we move through life, more time and events are continuously added—there is no bus stop at *Now*—the trip goes on and on, but unless we have a clock, we cannot mark the events by specific dates.

Therefore, we have no time, only an accumulation of events. This is another way of stating that time is not an entity in itself.

But still, there's no getting around the fact that time affects us on the deepest level. Waking up after a nap you don't feel that time has passed until you check the watch and find out you have slept for two hours. The timekeepers in your cells, registering continuous cellular events, however, have kept on ticking and you are now two hours older. Aging is a cellular process, the result of a myriad of small events along the timeline.

The brain's *perception* of time, on the other hand, differs from the cellular clocks in your body. Our awareness of time is a feature of our conscious life. During sleep and anaesthesia, you do not sense the time-lapse. You have to be fully conscious to be able to perceive time. And how we perceive it depends on several factors.

Imagine you are driving your car downhill on an icy road. You proceed cautiously and slowly since you can feel the slippery surface. Behind a sharp bend, you suddenly see another car in the middle of the road and between you and the other car is a girl standing on the road. You press the brakes, but your car just keeps sliding down toward the girl and the other car. You try to steer away but to no avail. A collision seems unavoidable. Time is rushing by and the options fly through your brain: Honk to warn the girl! Use the handbrakes! Release the brakes to get steering ability! Finally, you release the brakes, and the situation plays out without any damage or injury when you manage to steer the car to the opposite side of the road.

There was very little time to act, but in your memory the scene seemed to go on for minutes instead of a few seconds. In moments of danger, our perception of time becomes dilated due to horror and fear so we can consider several options and make a choice between them. This psychological "time dilation" may be an evolutionary gain as an aid to survive in a critical situation, giving more subjective time to respond to a threat. Experiments show that watching a movie with scary segments makes the viewer overestimate the duration of the clips, while film scenes that invoke sadness give a more correct idea of their length. It has been suggested that fear causes a state of arousal in the brain structure of the amygdala increasing the rate of the internal clock, to induce a "rescue state" in a threatening situation.

The subjective sense of time thus varies with your emotional state, and also with age, body temperature, metabolic rate, and it is also influenced by your health. Stimulating drugs, like caffeine, amphetamine and thyroid hormones lead to an overestimation of time intervals—creating a stronger sense of presence in the current moment—while depressants can have the opposite effect.

As we grow older time seems to move faster. The child aged five lives in the present before she or he is aware of the passing of time. And at the age of ten, an ordinary day seems longer to the grandson than it does to his grandfather. There are several explanations for this based on neurophysiological studies. One of them explains the phenomenon with "neural adaptation" meaning that with age we repeatedly receive similar stimuli that are already mapped by the brain and is therefore now unnoticed. They don't become registered events on the timeline.

But how do we humans actually experience and register time in our minds? How is time represented in the brain?

In 2014, the Nobel Prize in physiology or medicine was awarded to a trio consisting of the British–American researcher John O'Keefe and the Norwegian couple May-Britt and Edward Moser for "their discoveries of cells that constitute a positioning system in the brain." By doing experiments on rats running through mazes, they saw that there are specific neurons in the Hippocampus and nearby regions of the brain that are always activated when the rat is in a specific location in the maze. These "place cells" thus represent an abstract map of the maze inside the brain.

How this map is created is not known in detail, but it seems clear that this information is used to code sequences of experiences with spatial information. These ordered sequences are formed by registering inputs from different sensor neurons that fire simultaneously, creating so-called episodical memories. In this way, we can connect events and places as well as their order in time. By having the rats running in mazes where timing is important to get the reward, researchers have started to identify "time cells," neurons that fire at certain instances in time or with certain time intervals, which could give the episodical memories time stamps. This research is ongoing, but it seems like there is no dedicated clock function in the brain but regions with "place cells" can also contain timing information.

For physicists, time can be regarded as a continuous parameter (t) that enters all models and equations describing any flow of events. It is important, however, to realize that the only important notions are *differences* in time, $\Delta t = t_2 - t_1$, as in finishing time (t_2) minus starting time (t_1), measured by a stopwatch. Clocking Usain Bolt setting the world record $\Delta t = 9.58$ seconds over 100 meters in 2009, the velocity was $100/9.58 = 10.44$ meters per second, independently from time itself.

So, in the end, the ticking of an accurate clock represents our only objective measure of the flow of events, and we identify it with the flow of time. And since the flow of events always goes forward, almost by definition the flow of time goes forward. The arrow of time points from the past to now and then further into the future.

We never see the natural flow of events in reverse. The flames and smoke from a burning fire can never end up as solid pieces of fresh firewood. Nor can water streaming out of a faucet, reverse and flow back into the tap. Even if one plays a movie backward, so that persons and vehicles move in the wrong direction, and the hands of a clock in the film are turning anti-clockwise, the clock-time in the room of the movie projector is moving forward.

The arrow of time is always pointing from the left to the right in the timeline we created in the first paragraph of this chapter. It is always pointing in the direction of irreversible sequences of events that cannot be played backward. So, we say that time has no absolute meaning in itself. But in the same breath, we claim that the arrow of time always moves in the same direction. How do we unite those two ideas? We do it by using the concept of *entropy*.

The prevailing tendency in the world, throughout the universe, is to even out differences—in temperature, motion, and structure. An object that is warmer than its surroundings cools if left alone. A lump of sugar in a glass of warm water dissolves and the sugar molecules are spread out uniformly in the liquid. What is built up will always erode with time. In a few billion years, even the sun has gone out and become as cold as the space surrounding it.

This relentless drive toward equalization is the very definition of the direction of time. A hot glass of water does not spontaneously get warmer, only colder. The sugar molecules dissolved in water never reorganize themselves into a cube. If the glass falls to the floor, it shatters into pieces. The events never follow in the opposite order.

The overall tendency is that the disorder becomes greater—we say that *entropy* is constantly increasing.

Entropy is a classical concept of thermodynamics that was developed during the latter half of the 19th century when scientists and engineers were getting a deeper understanding of what was happening in steam engines. Entropy means "energy transformation" and was used to describe thermodynamic relations between energy, volume, heat, and work in terms of pressure, temperature, and chemical potential.

The direction of the arrow of time is linked to the natural flow of *entropy* in systems that are *out of equilibrium*. A system *in* equilibrium is by definition in a steady state—nothing happens to it. It has no evolution and therefore it is not dependent on time. It is the cold glass of water with all the sugar molecules dissolved. It is a burnt-out star in the cold universe. In everyday terms, it could be described as "dead." Nothing more will happen if it is left alone. It is a system that has reached maximum entropy.

In a system that is *out of* equilibrium, the entropy increases all the time. Until, with time, it arrives at a steady state, where nothing happens anymore. A building falls to ruins, a body dies and decomposes, a star burns out. That's how the arrow of time manifests itself.

But—*and here is the main point of this digression into time*—life itself seems to defy the laws of thermodynamics and the natural flow of entropy.

Every organism is in itself a complex system that is thoroughly out of equilibrium. And all the living plants, animals and bacteria on the planet are parts of a huge system of immense complexity that keeps on moving, breathing, and multiplying—rebuilding itself and even evolving. After almost four billion years, life is still around in full swing. It defies the drive toward higher entropy. But without the driving force of the sun, Earth's ecosystems would soon become a static system in equilibrium where entropy has reached its maximum, a dead world.

Actually, the total entropy of our planet keeps increasing. But the sun organizes at the same time life on Earth via photosynthesis and respiration. Energy keeps pouring in as long as the fusion of hydrogen atoms will power the sun. So locally, within the boundaries of living systems, the entropy *decreases*.

It was Erwin Schrödinger who introduced the concept of "negentropy" or rather entropy with a negative sign, in 1944 in his book *What is Life?*. He also commented later in the book that a more appropriate word for a physicist would have been "free energy," but he chose to call it "entropy with a negative sign" to make the point that living systems can lower their entropy.

So, life keeps on building ever more intricate systems despite the laws of thermodynamics, in thousands and thousands of ways. A special case in point is directly associated with the flow of time: the evolution of the circadian clock, a biochemical system that regulates timekeeping in many animals, insects, and plants on a 24-hour cycle—synchronized with the sun. It consists of an *oscillating biochemical reaction*—a kind of chemical pendulum—and a *genetic control system*.

Circadian rhythms can be beneficial to an organism in several ways. They help flowers to open and close at the right time. Nocturnal animals know when to lay low to avoid predators, photosynthetic bacteria can adapt their metabolism to the motions of the sun, and so on. So, it is no surprise that an effective system to keep track of time evolved early in the history of life on Earth.

A piece of this control mechanism was discovered by Stanford University scientists Ron Konopka and Seymour Benzer in 1971, identifying a specific gene—named "period"—in a mutant fruit fly. During the following decades, a complete gene-regulation network for the control of cellular timekeeping was identified. However, biological clocks must have existed long before the evolution of life had time to develop complicated gene networks to run the clocks. Most likely, periodic day/night variations gave rise to prebiotic chemistry capable of oscillatory behavior. So, where was the fundamental biochemical oscillator?

This early chemical pendulum turned out to be closely connected to a process that changed the Earth profoundly and created the opportunity for higher life forms to evolve—the Great Oxidation Event. In the beginning, there was very little oxygen in Earth's atmosphere, and the initial life forms were anaerobic, feeding on metabolic processes not involving oxygen. This began to change some billions of years ago mainly through the emergence of light-harvesting bacteria. These marine organisms began to manufacture oxygen in great quantities by the method we describe in Chapter 10—

photosynthesis. About 2.5 billion years ago, they had pumped out so much oxygen that the composition of the atmosphere had changed to become so rich in oxygen that it was toxic to the old anaerobes, but it paved the way for a new kind of life on a higher energy level.

But oxygen is a risky element. It is very reactive, ready to engage in fast energetic chemical processes that can be very destructive. (A drastic but familiar example is forest fires.) One way for oxygen to wreak havoc is through the formation of so-called reactive oxygen species—small molecules like hydrogen peroxide (H_2O_2)—not just a common hair bleach, but also a potent disinfectant; it kills bacteria. The early cells had to learn to handle these reactive molecules and to detoxify them in order to thrive in the oxygen-rich atmosphere. They developed a protective system in which certain proteins could neutralize the reactive molecules through a *redox system*. The proteins in question ≈ named PRX, *peroxiredoxins*—work by moving hydrogen and oxygen atoms around and, for instance, turn harmful hydrogen peroxide into harmless water—H_2O.

These antioxidant proteins created a kind of chemical dance, a self-oscillating reaction, where the levels of protein content vary back and forth during the day—eventually, this oscillation evolved to conform to the movement of the sun, probably following the natural flow of the photosynthesis of the plants. Much of the protein is available during the day when the oxygen flow is stronger, and this back-and-forth variation becomes a slow, but regular pendulum in the cells.

The discovery of the PRX-timer came in 2011 when two scientists in Cambridge in England, O'Neill and Reddy, found them in human blood cells. Those cells have no genes, but they still kept time in 24-hour cycles. The same basic redox oscillator was then also discovered in many other cell types, so redox oscillators seem to be the "grandfather clock," while the genetic feedback circuits is a later evolutionary invention that connects the clock to the surrounding cell tissue to control physiology and behavior.

However, the 24-hour circadian rhythm must be the result of a gradual adaptation through the eons. In the beginning, the Earth rotated so much faster that 3.5 billion years ago a day and a night lasted only 11 hours, not 24.

So, life organizes and orders molecular structures and lowers entropy, and thus seemingly defies the laws of nature. But it requires a living system to constantly receive new energy to maintain its high degree of organization. Life's processes are constantly being driven after more than three billion years with the sun as the ultimate engine.

The closest analogy to this phenomenon is a solar cell that charges a battery, so electric current can be stored or circulated to do work in various machines. Biological living systems make use of photosynthesis—photons from the sun to create energetic electrons in biomolecular photocells. Those electrons are then circulated in the system, feeding biological machines that create chemical "voltage" differences much like charging a battery, as well as creating energetic molecules that fuel the living processes. We will show you exactly how in Chapter 11.

Chapter 8

The Art of Finding Your Way Home

The power that makes grass grow, fruit ripen, and guides the bird in flight is in us all.

—Anzia Yezierska, an American novelist born in Poland

On a hot day in July 2004, an elderly man called the offices of a local newspaper in Malmö, in the south of Sweden. He asked if the paper had heard anything about ring-marked pigeons that might have appeared in the area. They hadn't, so the man told the reporter who took the call that the Racing Pigeon Club just had released two thousand young pigeons for their first competition in a town about 100 kilometers to the north. For a pigeon flying at a cruising speed of about 80 kilometers per hour, that would be a short flight of just under two hours. But three days after the start, only a quarter of the birds had returned. The rest, nearly one thousand five hundred racing pigeons, were gone. Pigeon breeders are well aware of the risk of losing a few birds in a race. But now three out of four had gone missing. It was an extreme case. The reporter who took the call knew nothing about the birds but decided to make a short news item about it. It was after all the middle of the slow summer month.

The news piece about the lost birds was published and the next morning the phones started ringing. Sure, people had seen pigeons. From all around the southern part of Sweden came reports of friendly birds with bands around their legs, happy to be fed. Some

Quantum Physics and Life
Ingemar Ernberg, Göran Johansson, Tomas Lindblad, Joar Svanvik, and Göran Wendin
Copyright © 2023 Jenny Stanford Publishing Pte. Ltd.
ISBN 978-981-4968-28-7 (Hardcover), 978-1-003-31267-3 (eBook)
www.jennystanford.com

pigeons were found in Denmark, and one had landed in Norway. The members of the Racing Club crisscrossed the region to pick up birds that had ended up at the wrong address. Some of them had settled on the newly harvested wheat fields where there was plenty of food in the form of waste-grain. Others never returned home. Instead, they joined the flocks of urban pigeons and made themselves a new life in the streets, far from the world of racing. But after a few weeks, most of them had found their way back home. However, the question of what had happened to the birds was still unanswered. Why did they disperse over such a vast area when they were supposed to fly straight to the coop? There was something a little spooky about the phenomenon. Was it the weather, or maybe the new 3G mobile phone transmitters that were being installed everywhere that had messed with their sense of orientation?

The theories were many. And the explanation, when it came, was actually extraterrestrial, but perfectly natural. Just before July 17, 2004, the day of the race, the sun had one of its recurring outbreaks—a flare. But this one was bigger than usual. A pulse of charged particles and energetic radiation was hurled through space at high speed and within a day's time, it hit the Earth. The result was a magnetic storm, a disturbance in the Earth's magnetic field that blew the birds off course almost like a hurricane. The pressure from the solar wind changed the very shape of the magnetic field. The field lines extending from the north pole to the south pole became displaced. And this, in turn, seemed to affect how the birds perceived their surroundings.

The fact that pigeons and other birds use the Earth's magnetic field as an instrument for finding their way during flights has been known for a long time. But exactly how they do it is still partially unclear. They have a sense that we lack, an ability to perceive magnetic fields. It is called magnetoreception and it appears as if they could actually see the magnetic field that surrounds us. The Earth's magnetic field is weak—it has a strength of 50 microtesla. That is about as much as the field next to a switched-on microwave oven—something we don't feel. An ordinary small refrigerator magnet has a field strength a thousand times that of the Earth. A magnetic resonance camera used in hospital examinations, on the other hand, uses a magnetic field that is 20 000 times stronger than the Earth's, but not even this powerful field can be felt by the patient. We simply lack a sense of magnetoreception. But other animal species seem to have one.

A number of famous experiments have shown that sea turtles and several bird species find their way when migrating with the help of the direction and strength of magnetic fields. The question is how it works.

A sensory organ for magnetism *could* be constructed with a magnetic material somewhere in the body, and that material might produce nerve signals to the brain. And, true enough, there are small crystals of the mineral magnetite in the beak of many migratory birds. Magnetite is a form of iron oxide found in many living beings. Bacteria, insects, and some rodents carry the mineral. We also have magnetite in our brains, with an unknown function. It doesn't seem to give us a built-in compass, though. But that is precisely what many birds have. Experiments have shown how signals from the pigeon's beak are sent via the so-called trigeminal nerve to its brain when the magnetic field around the bird changes. But that doesn't explain the whole process. On the one hand, it is difficult to understand how this mechanism could become so sensitive that it can provide the birds with enough information. Close studies of pigeons have also shown that their brains receive signals of magnetic changes via a number of other nervous pathways, signals with an unknown origin.

A fuller description of how the art of perceiving magnetic fields might work was presented by the German scientist Klaus Schulten as early as 1978. This description is firmly based on quantum physics. The traditional view is, as we have seen, that quantum phenomena occur only at low temperatures and in extremely short time frames. In the warm chaos of a living cell, quantum mechanical processes are expected to collapse within a few nanoseconds. Therefore, Schulten found it difficult to be taken seriously when he first presented his idea. Together with two colleagues, he wrote a paper explaining how a magnetoreceptive organ could work with the help of quantum mechanics. The influential American journal *Science* rejected the article with the comment that a more modest researcher might have put this idea into the trash can. Instead, the article ended up in a much less prestigious German journal of physical research. But the model they described in the article has since convinced many others.

Schulten got his idea after he realized that chemical reactions can be affected by weak magnetic forces. The principle behind this effect is based on the quantum mechanical phenomenon called entanglement and it works when two electrons are linked together to form an entangled pair, a phenomenon we described in Chapter 4.

The mechanism for magnetoreception that Schulten proposed looks something like this: Electrons normally move in pairs, two and two, around a molecule. Sometimes this dancing couple can be pried apart by a particle of light, a photon. The photon gives one of the electrons an energy kick and sends it over to a neighboring molecule. The result is a pair of two molecules where one of them lacks an electron—it has a "hole"—while the other has been given one electron extra. The molecules are now sharing an electron-hole pair and have become what you call a "pair of radicals."

In such a radical pair, the molecules constitute a common "two-electron state" where they can react to variations of the magnetic field lines. Changes in the inclination of a surrounding magnetic field will give the pair slightly different chemical properties. This sensitivity depends on the quantum mechanical spin of the two electrons in their new state. Here Schulten saw the possibility of how a high-sensitivity sensor for changes in weak magnetic fields could work. But no one had seen it in the wild.

Since such a mechanism requires light, photons, to function, it was natural to imagine that it would be situated in the eye. A few years after the idea was presented, the protein cryptochrome was discovered, a substance found in the retina of many animals. Cryptochrome is sensitive to light and proved to act as a regulator of the organism's internal clock—for example, to set the daily rhythm. This circadian rhythm is naturally governed by changes in the light, from night to day. The sensitivity of the cryptochrome comes from its ability to change chemically by forming radical pairs when exposed to blue light. When Schulten realized this, he and his colleague Thorsten Ritz worked out a more detailed description of what the bird's internal compass could look like: it probably sits in the eye and is based on cryptochrome. Ritz and Schulten assumed that the cryptochrome molecules are placed in parallel to one another on the retina. Since the retina is a curved, concave surface, a weak magnetic field would fall toward the photosensitive molecules at different angles depending on where they are on the retinal surface. The result would thus create a pattern on the retina, an image of the magnetic field. In other words, it is conceivable that the Earth's magnetic field creates a visual impression in the bird's brain, and that the inclination of the field lines gives a sense of orientation.

A "bird's-eye view" is an old metaphor for looking at the world from above. But in nature, it might mean more than that. It could mean a world built on completely different impressions than the ones that we get. We know that birds can see ultraviolet light and the way that light is polarized. In addition, they probably see the Earth's magnetic field. Whether it is in the form of magnetic lines or color changes we do not know. But we know that the migratory birds have a magnetic sensing system that directs their beaks toward the distant goal: a place on Earth where there is hope of food and where the chances of survival and reproduction are good.

Schulten's and Ritz' hypothesis is promising, but there are objections. Is the Earth's magnetic field strong enough to be registered in this way? And is the quantum state, that is, the entanglement of the electrons in the cryptochrome radicals, sufficiently long-lasting for the mechanism to function. It has been suggested that the spin coherence time must be at least one microsecond for the weak geomagnetic field to impact the system. There is a strong indication that this is so, but it is not yet accepted by all researchers in the field. Many animals walk, swim or fly unbelievably long distances to find food or to mate. We know for a fact that Earth's magnetism affects the long journeys of sea turtles. Insects also move long distances.

The bogong butterfly lives and breeds in the plains of south-eastern Australia, but each spring millions of these nocturnal butterflies fly to spend the winter in caves in mountainous regions one thousand miles away. We don't know how they find their way there, but a strong hypothesis is that they navigate with the help of cryptochrome in their antennas that senses the Earth's magnetism. The human retina also carries some cryptochrome. Does that mean we have a rudimentary ability to feel—or see—magnetic fields? If you give the human gene for cryptochrome to a fruit fly, it will react to weak magnetic variations. So, our version of cryptochrome does work as a magnetic sensor in an insect.

Whether magnetoreception can help humans to find their way home is more questionable.

Chapter 9

The Vision in New Light

One way to open your eyes is to ask yourself, What if I had never seen this before?
What if I knew I would never see it again?

—Rachel Carson (2011). *The Sense of Wonder*, p. 24,
Open Road Media.

During the fall of 1940, the future of Europe was in the balance. German troops had occupied France and were ready to invade Britain. A defeated England would let the German armed forces concentrate on the attack against the Soviet Union, Hitler's most important aim for the war—to create *lebensraum* and a German colonial empire in the east. With a victory in the west, all forces could be deployed against the Russians.

The German intention was to quickly knock out the British Air Force, to open the way for an invasion over the channel from France. But the plan failed. The aerial war that came to be known as the Battle of Britain became a victory for British aviation—"Never was so much owed by so many to so few," as Prime Minister Winston Churchill famously described the outcome afterward.

Germany then changed tactics. Instead of an invasion, England would be bombed into submission. A wave of terror bombings lasted for more than six months from the fall of 1940 to May 1941. Tens of thousands were killed by the carpet bombing, entire neighborhoods

Quantum Physics and Life
Ingemar Ernberg, Göran Johansson, Tomas Lindblad, Joar Svanvik, and Göran Wendin
Copyright © 2023 Jenny Stanford Publishing Pte. Ltd.
ISBN 978-981-4968-28-7 (Hardcover), 978-1-003-31267-3 (eBook)
www.jennystanford.com

were smashed to pieces, and whole industries were destroyed. London alone was attacked more than 70 times. The British Air Force had a hard time combating the armadas of German bombers that swept in under cover of darkness. During the first months, the British fighter planes managed to stop only a few percent of the German bombers. But the odds would change. During the spring months of 1941, RAF pilots became better at detecting and shooting down the Germans. The reason behind the higher rate of success, according to the British authorities, was skilled pilots with a first-class ability to see in the dark. And sharpest of them all was a young man named John Cunningham, nicknamed "Cat's Eyes."

Cunningham shot down at least 15 German planes during the early months of 1941 in his Bristol Beaufighter. And his good night vision came from eating lots of carrots. At least that was what the propaganda department of the British War Ministry wanted to make the Germans believe. In fact, Cunningham was one of the first pilots to be equipped with a new lightweight radar that could be placed on board an aircraft. The pilots no longer had to wait for radio instructions from the ground about the location of enemy planes; they could find the targets themselves.

Carrots and other root vegetables were cheap and available domestic foodstuffs, during the wartime rationings. The only difficulty was to make people eat more of them. So, the carrot was presented as an excellent source of nutrition that also helped the war effort in a very direct way. The heroic pilots of the RAF used them to stop the night bombers, and the propaganda story of the carrots also helped to hide the existence of the new secret radar equipment.

We don't know whether the Germans really bought the story of the carrots, but it was nevertheless considered to be good for the home front. Even when there was a shortage of other groceries, there were plenty of carrots. Hitler decided in the spring of 1941 to attack the Soviet Union without first defeating England. Thus, most of the German Air Force was transferred to the Eastern Front and the bombing of England ceased. But the myth of the carrots lived on. Even today it's a common claim that carrots give you better night vision.

Carrots are important to vision, but perhaps not so dramatically important that they weighed in as Europe's future hung in the balance in the 1940s. To get a better look at the role carrots really play, we need to approach the innermost parts of the eye from a quantum physical perspective. Eyes are fine-tuned instruments with lens, aperture and sensitive receptors linked to an advanced brain imaging array via a system of nerve cells. You can lead a good life without the information we receive from the light—visually impaired people do it every day. But it is difficult to imagine what the world would look like if we didn't use this information at all.

Eyes of various kinds have existed in the animal kingdom since the first photosensitive cells emerged during the Cambrian period 540 million years ago, when multicellular life arose in the oceans.

There are many kinds of eyes. Mussels have mirrors instead of lenses in their eyes to see the world a little sharper. Insects have facet eyes that consist of hundreds of hexagonal lenses. The mantis shrimp is considered to have the most advanced eye system on Earth. On two antennas it has a number of constantly moving facet eyes. With them, the predatory shrimp can see the polarization of light, infrared, and ultraviolet light, and detect the slightest movement in the environment. But no matter how advanced the eyes are, animals and humans have access only to a narrow part of the entire electromagnetic spectrum—at best approximately 300 to 800 nanometers in wavelength. These are the colors of the rainbow, extending into infrared and ultraviolet.

All animals with eyes carry the same genetic key sequence that controls the development of the visual organs. The Swiss biologist Walter Gehring has mapped out much of what we know about how genes shape an organism during fetal development, from a simple embryo to an adult organism. He has shown that there is a gene that goes by the name Pax 6 which acts as the main switch for the development of visual organs and the production of photosensitive proteins, the so-called opsins. Pax 6 initiates the construction of eyes in a fruit fly as well as in a mouse and humans. And if you transfer a Pax gene from a fly to a mouse, it still works; the eyes develop normally in the mouse. This means that the mechanism came into existence very long ago in the history of evolution and has been preserved almost unaltered.

A few years ago, Gehring found an almost identical Pax gene in a group of jellyfish, an animal that belongs to the most primitive forms of multicellular creatures. These genes, in turn, may have their origin in photosensitive algae, so-called phytoplankton, that make a living from sunlight through photosynthesis. Sometime in the early stages of life, such algae may have been picked up by jellyfish and become part of them in the kind of biological cohabitation that is called symbiosis. Variants of this genetic apparatus have since been inherited through millions of years, and slightly modified ended up in us. What was once a tool for single-celled algae to capture energy from the sun has thus become the basis for our visual ability. It just goes to show how conservative evolution can be. Things that work will be used over and over again. It is also proof of how all living things are connected. The jellyfish that once incorporated the photosensitive algae in their squashy bodies are our distant relatives.

The parts of the eyes we show the world are sometimes called the "mirror of the soul." But what we see are only the parts that regulate the inflow of light: cornea, iris, and pupil. The eyes are much more than that. It is on the inner, posterior wall of the eye, the retina, that the visual impressions themselves begin to form. Light-sensitive cells in the retina react when visible-light photons hit them, creating nerve impulses.

The nerve signals are transmitted to the cerebral cortex where they are transformed into images. Parts of this constant stream of photons become conscious impressions, others just pass us by. Some end up as stored memories, but most of them are simply forgotten. The flow of information from the eye is made possible by quantum physical processes in the eye. And those processes are extremely sensitive.

The retina consists of three layers of cells: photosensitive receptors, a layer of so-called bipolar cells and a layer of nerve cells. The light receptors are at the back of the retina and consist of two types of photosensitive cells, rods and cones. There are 120 million rods in each eye, and they cannot perceive colors. However, they are effective in low light. The color vision is in the cones.

It is here, in rods and cones, that one finds the photosensitive opsins that we seem to have inherited from the sea algae and jellyfish. When the light of right wavelength hits a rod or a cone, information is sent to the next cell layer, the bipolar cells, which in turn send a signal to the nerve cells and further on through the optic nerve to the brain. In this three-step process, the opsins play a key role. They are photosensitive molecules, proteins that have been hanging around ever since single-celled organisms in the Cambrian seas began to distinguish light and darkness hundreds of millions of years ago. The opsins are molecular structures made up of hundreds of atoms, microscopic machines tailored for a single-precision task: to respond to light. Like all machines, they consist of several parts, and in a pocket between the large arms of the protein molecule lies the key to our visual ability—the part of the opsin called retinal.

When we look at the world around us, photons flow into the eye and hit the millions of photosensitive cells in the retina. When a photon reaches the photosensitive protein—the opsin—it is picked up by the retinal part. This means that the protein molecule gets an energy boost—it gets excited to a higher quantum level and changes its three-dimensional shape. It becomes activated.

The retinal shape change causes a shift in the electrical balance of the molecule—the cloud of electrons surrounding the protein becomes displaced. This electrical redistribution then creates an impulse that is transmitted to the optic nerve's main cable toward the brain's visual center: part of a visual impression is on its way.

After these lightning-fast chemical events, the activated cells return to the start position after about 30 minutes to receive new photons. This is the time it takes to get used to darkness or to regain full vision after being blinded.

The reaction when the retinal is charged with the energy of a photon and starts the nerve signal is one of the fastest and most effective light reactions in nature. The explanation lies in the eye's ability to exploit the possibilities of quantum mechanics. There are cases where single light particles are enough to affect large living systems. Be it a clam digging down to hide in the sand, or a dung beetle using the light from the Milky Way to find its right path. There is much evidence that this hyper-sensitive way of extracting information from light is an extremely ancient phenomenon. We are talking about hundreds of millions of years.

Photons can be described as particles of light. But in the quantum world, they are—as we mentioned—at the same time also waves with the properties of waves. They are thus delocalized (see Chapter 4). Photons in the visible region are delocalized over hundreds of nanometers, which means that a photon can simultaneously affect many opsin molecules without having to jump around. But quantum mechanically, the photon is eventually absorbed, within a few femtoseconds, by the retinal in a specific opsin molecule, creating an electron excitation that causes the retinal molecule to change shape in picoseconds. The time from the light hitting the retina's rods and cones until the signal is sent to the nerve cells is therefore counted in picoseconds.

The fact that the retinal molecule can change its shape, *isomerize*, after absorbing only a single photon at the right frequency makes the human eye one of the fastest and most sensitive light detectors we know of. Thus, in some cases, single light particles are enough to affect large living systems.

To see in the dark, we usually need night-vision devices that can detect infrared radiation from warm bodies. However, recently patients undergoing so-called photodynamic therapy, PTD, to eliminate cancer tumors also somehow got to see better in the dark, an unexpected bonus effect. The explanation is tied to the quantum spin of oxygen. In this treatment, a light-sensitive molecule—a photosensitizer—is excited by laser light, and the excitation is then transferred to an oxygen molecule, O_2, exciting it from the ground state to its first excited "spin-singlet" state. Singlet oxygen is an energetic little "bomb" that can damage biomolecules when it gets in contact with them. That is why it is used to damage and kill cancer cells.

The photosensitizer used is a molecule of chlorin. The surprising result of the PTD cancer treatment was that the infrared-sensitive chlorin molecules were also attached to the opsin proteins in the eye. And when a chlorin molecule was excited by a single infrared photon at night, the excitation created a singlet oxygen molecule that would trigger the shape change of the retinal molecule.

So, one could say that the retinal molecules had been equipped with night-vision goggles that made it possible for a single infrared photon to indirectly trigger the change in the retinal structure that is usually "designed" to be triggered directly by visible light.

It is obvious that retinal is a substance that is absolutely crucial for our ability to see light. But still, it cannot be produced by the body. We have no gene for making retinal, it must be added from outside. Retinal is a form of vitamin A that is formed in the body with other types of vitamin A as raw material, one of them being the beta carotene that gives carrots their characteristic color.

Deficiency of vitamin A is rarely a problem for people in affluent parts of the world, but it is one of the most important causes of high infant mortality in poorer countries. It can also, for obvious reasons, lead to deficient eyesight and even blindness. However, we do not get better eyesight or night vision the more carrots we eat. For those who want to see things that hide in dark, radar can be a more important tool, as it was for John Cunningham and the RAF 1941. But don't underestimate the humble carrot. Created by photosynthesis and supplier of essential vitamins—one of our many partners in life's almost endless variety of creative processes.

Chapter 10

The Photosynthesis and the Golf Putt

The grass is spreading out across the plain. Each year, it dies, then flourishes again.

—Bai Juyi

Albert Einstein was once asked if he played golf. He replied that he had tried, but thought the game was too difficult. Einstein's only documented moment on the golf course happened because Abraham Flexner, the man who founded and led the Princeton Institute of Advanced Study, where Einstein stayed in the 1930s, was a dedicated golfer. Flexner's repeated invitations made Einstein finally take a golf lesson from a young coach. But the famous professor was obviously not disposed for ball games. After several unsuccessful putts and almost as many good tips on posture, looks, breathing and everything else needed to make a good putt, Einstein had had enough. He set the club aside, asked to get four golf balls, shouted "catch!" and threw all four balls at the coach. The coach failed to catch a single ball.

"Young man," Einstein said, if I throw a single ball, you can catch it. When I throw all four you catch none. The same thing goes for teaching. Explain one thing at a time!

The story may not be entirely true, but it was retold by a person who met one of those present when it allegedly happened. Nevertheless, the story is so good that it can serve as a simple

Quantum Physics and Life
Ingemar Ernberg, Göran Johansson, Tomas Lindblad, Joar Svanvik, and Göran Wendin
Copyright © 2023 Jenny Stanford Publishing Pte. Ltd.
ISBN 978-981-4968-28-7 (Hardcover), 978-1-003-31267-3 (eBook)
www.jennystanford.com

introduction to one of biology's most important phenomena: photosynthesis. The basis of (almost) all life on Earth.

There are only two original sources of energy on Earth driving natural life. One is the heat from Earth's molten core. The other is the sun, our own star. Eighty percent of all the energy we use is solar energy in the form of fossil fuels stored since hundreds of millions of years as coal, oil or gas; conserved solar radiation bound to matter through photosynthesis, a process that utilizes quantum physics to function.

Photosynthesis is the most important single biological process we know of. It transforms the vast flow of photons from the solar fusion reactor into chemical energy, which in its turn is the engine of all life on Earth. It is also photosynthesis that creates the vital oxygen in the Earth's atmosphere.

As early as the 19th century it was clear that plants grow by binding carbon dioxide and water—so called carbon fixation—releasing oxygen and that they need sunlight to do so. The term photosynthesis simply means that the plant produces (synthesizes) carbohydrates using light.

Already the first fossil cells from 3.5 billion years ago may have lived by harvesting solar energy. Between 3 and 2.5 billion years ago, organisms evolved that produced oxygen using photosynthesis, and this oxygen slowly filled the atmosphere and changed the composition of the air. Sometime between 1 billion and 500 million years ago, there was enough oxygen around the Earth to form the ozone layer in the atmosphere that protects life on land from the harmful ultraviolet radiation from the sun.

Photosynthesis means that water and the carbon dioxide in the air are transformed into complex organic molecules that build all plants, from grass to roses, cacti, pines, and apple trees. Scientists have been trying to imitate this process for decades, but so far with limited success. It is not only that photosynthesis utilizes sunlight super-efficiently. It produces useful energetic molecules from the simplest and cheapest possible raw materials: sunlight, water, and carbon dioxide. It drives the cellular machineries that create fruits and vegetables, food for billions, fuel, building materials, drugs and much more.

The efficiency, the ability to harness the sun's energy, is astoundingly high; over 90 percent of the captured solar energy is utilized. The very best photovoltaic cells produced today have an efficiency of around 30 percent. We now know that different photosynthesizing species—all green plants and some bacteria—have developed slightly different variants of the method of capturing the Sun's energy. But they all use the same basic first step to convert the energy from the photons of the sun to electronic excitations in the small beads inside the plant cells called chloroplasts.

The solar energy is captured with "antennas" inside the chloroplasts, which consist of a few hundred molecules of chlorophyll or carotene. The antenna molecules absorb light with specific wavelengths. Chlorophyll is particularly effective at capturing blue and red light while reflecting green, which means that plants are mostly green. Carotenoids reflect in the red-yellow area, and this is what we see in the fall when the leaf's green chlorophyll breaks down.

Located within the ring of antennas are proteins that form a reaction center. This is where the energy from the sunlight is used to create adenosine triphosphate (ATP), which is the universal fuel in all living organisms. The solar energy is then stored as an energy supply in the form of ATP molecules, to be used for all energy needs of the plant or the bacterium.

When the sun shines on a green leaf, it is only the green chlorophyll of the antennas and the other pigment molecules that capture the photons. By using the energy in a photon, an electron in the "antenna" is raised to a higher energy level. It gets excited and jumps up a step with the help of the photon. And on the level where the electron was before there is now one electron less—a hole. Thus, the solar energy is temporarily trapped in an excited electron-hole pair, a so-called exciton. Now it is important to transport that exciton to a reaction center where its package of energy can be converted into fuel.

The whole complex of proteins that the energetic exciton particle must go through to find the right address is microscopically small from our perspective but gigantic from an atomic point of view. A single electron is here a bit like a tourist in an unknown metropolis trying to find his hotel by randomly running along the streets. In the end, it might work, but it may take a long time to find the right address. And it is likely that those who try, lose their energy on the way. So, how do they do it?

This is where the golf course comes in as a good metaphor. In photosynthesis, the incoming photon creates a particle (a golf ball) in motion, supposed to find its way to the reaction center (hole). What is so fascinating is that the photon almost always, at least 99 times out of 100, makes a hole-in-one. And this even though the "green" is in reality a large biological molecular complex at room temperature, which in aqueous solution is also constantly changing. In other words, it moves. Also, every photon is new and completely untrained. It is the first time it has tried to hit the hole and there is no instructor to ask for advice. Thus, the golf green of photosynthesis must be very special that can make any player succeed. Maybe even an impatient professor like Einstein.

What has fascinated scientists is that the exciton's journey from the pigment in the antenna to the reaction center is so efficient. In principle, all the absorbed photons reach the target. Even though the excitons have less than a nanosecond to get there. So, the exciton has to reach its target much faster than that. If it doesn't, the solar energy is lost, radiating away as heat, to no use. And since we know that plants can handle this over and over again, and with extremely high efficiency, we can assume that nature has found a smart solution.

The pathway through the antenna complex passes through a series of large pigment molecules down to the reaction center. The exciton thus needs to find its way from pigment molecule to pigment molecule, which is thought to happen through random jumps. This random walk should preferably lead to the excitation finally reaching the reaction center, where the energy can be used to create the charged ions (electrons and protons) which are then used to produce the fuel ATP.

The most well-studied reaction is found in a kind of green sulfur bacteria that lives in very dark environments. These are thought to be similar to the first organisms three billion years ago that harvested sunlight without producing oxygen. Before 2007, the common understanding was that photon from the sun is captured by the antenna and creates an exciton at the periphery of the reaction center. The exciton then hops down a sort of ladder toward the inner part of the reaction center. On the way down, it loses a little energy for each hop, until it finally delivers the remaining energy to its final destination.

This picture was challenged in 2007 when the group of Fleming at the University of California at Berkeley published a paper in the journal *Nature* with the title: "Evidence for wavelike energy transfer through quantum coherence in photosynthetic systems." The paper described an experiment where they used extremely short laser pulses to zap the antenna complex and then observed coherent oscillations on a picosecond time scale. The result suggested that the exciton moved as a coherent wave down the chain of pigments to the reaction center.

The idea of exciton waves with picosecond coherence times was Earth-shaking. Many groups jumped on the wagon, and during the following 10 years, physicists adopted and promoted the radically new view of quantum coherence over long times in warm and wet environments. So long that they would be at the core of photosynthesis.

In order for the quantum mechanical wave properties to matter, the particle must not be disturbed by other molecules that get in the way or by vibrations in the large network of proteins where the pigment molecules are suspended. It would disrupt the quantum mechanical state and create decoherence—break the wave. If this happens, the exciton wave slows down and eventually comes to a full stop on one of the pigment molecules along the way. And then there will be no photosynthesis. Since photosynthesis takes place at room temperature in biological molecules surrounded by a variety of water molecules, many scientists had previously not even thought of the idea that quantum mechanical wave properties could play a role in the green leaves of plants. Of course, it was considered that decoherence would occur in such a biological chaos. That was also why the image of the randomly jumping particle was the natural one. But it turned out that on these very short time scales, the chaotic environment simply cannot destroy all quantum mechanical wave effects.

In 2017, the empire struck back. A prominent international collaboration repeated the experiments with refined technology and published a paper with the title: "Nature does not rely on long-lived electronic quantum coherence for photosynthetic energy transfer." Other investigations concluded that oscillations of picosecond time scales were due to vibrational motion in the pigments.

The new results force us to conclude that, after all, the exciton is not propagating as a coherent wave to the center on a picosecond time scale. The results rather confirm the "good old" understanding that the exciton hops incoherently down the energy hill while losing a little energy in each hop by creating vibrations in the pigments. And the energy hill is so constructed that the exciton is precisely directed to the target—hopping along a chain of pigments leading directly to the core of the reaction center.

In a sense, it would have been more fun, at least for quantum physicists, if coherent wave-propagation had been the reason for the efficiency of the photosynthesis. But one can just as well turn this around: it is equally fantastic, and even more important, that robust *incoherent* quantum physics describes the engine for life on Earth.

Returning to the game of golf, the excitons become balls in what we might call a game of quantum golf, a game that doesn't quite work as the classic variant.

We already described how one can use modern laser technology to investigate chemical processes that take place during extremely short periods of time—we are talking about femtoseconds. This has made it possible to follow the path of the exciton ball toward the hole on the golf green of photosynthesis, although it moves pretty fast compared to a golf ball.

When you make mathematical models of how such an exciton transfer can work, you see that a balance is needed between the fast wave-like transport and the friction that can arise from the disturbances that occur on the road after all. In other words, it must not go too fast. Here again, photosynthesis works like golf. If the golfer strikes too hard, the ball will jump over the hole, even if the direction is exactly right. If the "friction" is instead too strong, then the ball will stop before it reaches the hole. The right balance is needed between the undamped motion of the ball, which corresponds to the wave movement of the exciton, and the friction (decoherence) given by the grass (the surroundings). In addition, the photosynthesis green must be bowl-shaped, so that the ball moves toward the hole regardless of in which direction the untrained photon strikes.

A complementary picture, that is closer to the reality of the quantum game, is to look at log-flume riding down the waterfalls at an amusement park. The exciton is then a hollow log with a happy family that floats between lakes of pigment proteins down waterfalls connecting the lakes. The levels of the lakes can represent the energy levels of the pigments, and the log falling down the waterfalls represent the tunneling of an electron hopping from one pigment to the next. When the log hits the next lake, there is a big splash and the log loses most of its speed—this is the effect of quantum friction. In this exciton quantum log-flume ride, the physical splash is a vibrational wave excited in the lake of the pigment and its surrounding landscape.

In the amusement park, the log is fortunately gently guided to the exit, to let the happy kids run on solid ground to the next event. In the quantum game, however, the next event is more dramatic: the exciton log falls downhill to a landscape of lower lakes and is eventually "pulverized," transformed into chemical energy used to drive the metabolic chemistry of living cells. In this way, the exciton is assisted by vibrations in the pigment molecules plus a good deal of attenuation from the surrounding protein network, to reach the reaction center 99 times out of 100.

It is obvious that Evolution has taken a long time to develop different photosynthetic complexes, and that the drive toward high efficiency has been strong. Researchers continue to study photosynthesis, not the least with the ambition to develop better artificial solar collectors that can help solve humanity's increasing energy needs. With ever better experimental methods and greater computational power, one day we will most certainly have a more accurate picture of how photosynthetic complexes work, and thus also the ability to imitate its efficient way of using the sun's energy. One question that then remains is whether it can help us to improve the golf putt.

Chapter 11

The Respiratory Chain Sustains Our Lives

Available energy is the
Main object at stake in the
Struggle for existence and
The evolution of the world.

—Ludwig Boltzmann

Staying alive is no free ride. The price to pay is energy. All living creatures spend energy to move, grow, or simply keep their structure intact. We all need to get it from somewhere. This is true for animals, plants, fungi, and bacteria.

It takes around 80 watts of energy just to keep our bodies in a resting state, but our bodies can increase the expenditure up to 1 000 watts during short energetic bursts of action. For activities that last a whole day the output is limited to around 300 watts. For comparison, a toaster preparing your breakfast toast uses about 950 watts, and a microwave oven about 100 watts for defrosting and up to 1 000 watts for cooking.

To keep functioning, we need to constantly add energy to our systems. Energy is also stored in cells in the form of chemical energy in carbohydrates, lipids, and proteins—the basic nutrients. All the fabrics of our bodies, like cell membranes, organelles, and enzymes are impossible to produce without the influx of energy, and these

Quantum Physics and Life
Ingemar Ernberg, Göran Johansson, Tomas Lindblad, Joar Svanvik, and Göran Wendin
Copyright © 2023 Jenny Stanford Publishing Pte. Ltd.
ISBN 978-981-4968-28-7 (Hardcover), 978-1-003-31267-3 (eBook)
www.jennystanford.com

structures hold stored energy. And all this energy ultimately comes from the sun.

The law of "conservation of energy" tells us that energy can be converted from one form to another, but never be created nor destroyed. It may have the form of kinetic energy of a moving object, or potential energy in a field of gravity, electric potential, or magnetism. It may also have the form of chemical energy, radiation energy or thermal energy. Through photosynthesis, the energy from the sun is converted into the stuff that plants are made of. Much of that stuff, carbohydrates, is food for other beings, who themselves can become food for creatures higher up in the food chain. Creatures like us.

So, an ordinary lunch might include a mixed green salad with a dressing as a starter. The salad is grown in a garden, fed with water and carbon dioxide that were transformed into carbohydrates with added energy from the sun through the process of photosynthesis, now ready for consumption. Solar energy in the form of photons is stored in the plant material, a lot of it edible—it depends on what kind of creature you are and how your digestive system functions. Salad works for us. A nice dressing adds some useful fatty acids and, of course, some flavor.

So, to the main dish—meat with potatoes. A beef cow has been grazing on grassy fields and its body has converted this simple fare into muscles and tissues. The animal was slaughtered, and the muscles turned into steak and prepared for ingestion. It becomes part of a delicious meal that adds energy to the body for chemical and mechanical work and the production of heat.

After a meal, the nutrients from the food—carbohydrates, proteins, and fat—begin their transformation in the digestive system. To prepare for the energy conversion, the food is first broken down into its parts. Carbohydrates become the simple sugar glucose, proteins are split up into their component amino acids and fat from the diet becomes lipids and short-chain fatty acids. The most important source of energy is glucose from carbohydrates. This chemical dismantling of the nutrients produces carbon dioxide—CO_2—as one of the waste products of the process.

Then starts an intricate process that sometimes is called the "citric acid cycle" or the "Krebs cycle." This happens in the tiny sausage-like bodies in our cells that go by the name of mitochondria.

They are the universal energy converters in all animals. The number of mitochondria in human cells vary. Red blood cells have no mitochondria while liver cells may have around one hundred of them. To see them, you need a good microscope; they are less than one-millionth of a meter—a micrometer—in length.

The degraded food molecules then release their store of chemical energy through a sequence of electron transfers. Electrons travel through a series of enzymes residing in the mitochondria. They are called complex I, II, III, and IV respectively. This is a transport chain, where electrons successively drop down to lower energy levels, one at a time, like falling down flights of stairs. Each step downwards releases a unit of energy, and that energy is used to pump ions of hydrogen—protons (H^+)—from inside the inner membrane of the mitochondria, to the outside. The resulting surplus of positively charged protons on one side creates an electrical potential across the membrane. This difference in electric charge, a gradient, becomes a source of energy, like a charged battery, that can drive vital reactions. The most important one is the formation of energy packages called ATP (adenosine triphosphate), the stuff that constitutes fuel for all living cells.

The whole process is called respiration. Simply put, it is photosynthesis driven backwards. Energy is harvested by converting carbohydrates like sugars and cellulose into carbon dioxide and water (although some energy is lost. No conversion process is perfect.)

The electrons that are transferred through the enzyme complexes in the mitochondria finally end up with a molecule of oxygen gas (O_2). The oxygen has traveled to this very site with the blood, delivered by a hemoglobin molecule. The addition of electrons makes the oxygen combine with protons (H^+) from the surrounding solution to make water—H_2O. This is the final destination and end terminal for the mobile electrons, and they are now at their lowest energy level; all the available energy has been extracted from the food, with carbon dioxide and water as the rest products.

This is the reason the body needs oxygen—if there is no oxygen the mitochondria will stop working—the respiratory chain will have no final electron acceptor, which means that the flow of electrons down the stairway dries up. Without oxygen as the terminal station for the electrons, energy production will cease. The molecular order arranged by life's molecules disappears and disorder, increasing entropy, takes over.

This is what happens in the heart after an infarction or in the brain after a stroke. Lack of oxygen results in cellular death, and this is of course the reason why we need to breathe. In the respiration process, nearly half of the energy content in the food can be extracted and made into ATP molecules. That's a decent level of efficiency compared to other conversion systems, like a car engine, where no more than around 30 percent of the chemical energy in the gasoline is transformed into mechanical work, and the rest is dissipated as heat.

The human body produces millions of ATP molecules every second through the respiration process, and these fuel units are constantly used and recycled. We all produce and consume our body weight in ATP every day.

Protons may also re-enter the mitochondria without forming ATP. This transforms the energy of the proton gradient into heat instead. That is what happens in brown fat and helps to keep the heat in small babies and hibernating animals like brown bears.

The mitochondria are independent bodies in the cells—small organs or organelles—with double membranes and their DNA. That means that they run their genetic program in parallel with the genome in the cell nucleus. Why is that? And where do they come from?

The dominant explanation today is that they are the distant relatives of bacteria that once found their way into cells a long time ago. The living creature that is genetically most similar to mitochondria is a kind of bacteria called rickettsia, but there are a number of other candidates. Much of their original structure is now lost, and mitochondria have evolved in many different ways in the animal world.

Whoever they were, these ancient bacteria stayed in the cell and lost all the structures they didn't need and kept just the basic energy conversion machinery—to our benefit. This happened at a time when the oxygen content in the Earth's atmosphere was increasing. The mitochondria made practical use of the abundant oxygen at the time when the life forms with cells lacking mitochondria all but disappeared. Oxygen was the future, and those with mitochondria could flourish.

In a similar fashion, the chloroplasts of the green plants, where the energy harvest of the photosynthesis takes place, are also cellular invaders like the mitochondria. They were originally green cyanobacteria, a kind of marine microalgae that made their way into plant cells.

The high efficiency of the electron transport between the electron donors and acceptors, the enzymes in the cycle, is rapid and efficient. It doesn't work the same way in all animal cells, but the basic process is the same, the one described here.

If we accept the theory that the mitochondria originally were free-living bacteria, that moved in with us when the atmosphere was being enriched with oxygen, then we can safely say that they have been working to convert energy in cells for more than three billion years. So, evolutionary pressure has had plenty of time to improve and perfect the workflow.

It was suggested already in 1967 by the two Americans Don De Vault and Britton Chance that quantum mechanical effects played an important role in how the electrons were juggled down the energy levels in the enzymes. If the electrons could tunnel through some of the passages along the way, that would be a possible explanation for the high output in the respiration cycle. Tunneling opens up ways to transfer the electrons with a minimum of energy losses on the way—ways that couldn't be described by classical physics.

It is not evident that such long electron transfer that is necessary for this model can happen via electron tunneling, but nowadays this is nevertheless a generally accepted hypothesis. Quantum physics is certainly at the basis of the two most important biochemical events on the planet—the harvesting of sunlight by the plants, and the next step in the food chain, the eating and biochemical processing of the plants by animals.

Life demands a continuous input of energy not only to enable mechanical work, heat, or electricity, but also for maintenance of the machine itself, to keep the body functioning through constant upkeep. Compare a muscle cell with a steam engine. If you don't burn any coal the steam engine will stop moving. If you do not supply the muscle cell with energy it will also stop moving—but it will also degenerate. No maintenance and entropy increase takes over and the cell disintegrates. If you leave the steam engine for a hundred years and then supply it with energy again, it will probably start moving its pistons and wheels—unless it's too rusty. Adding heat to a disintegrated muscle cell will not produce work but rather increase entropy—life is replaced by death. The route leading toward increasing entropy is much faster in us compared to a steam engine.

The input of energy passes through you and me to counteract entropy growth—keeping us alive. Then it moves on—converted but neither created nor destroyed—it is still around, perhaps for someone else to use.

Chapter 12

A Sense of Smell

Scent is the strongest tie to memory.

—Maggie Stiefvater

What do plants want? Of course, they have no will, at least not in our sense of the word, but they can really affect other organisms with a definite purpose. Plants can make other creatures help them with problems they are unable to solve themselves. They attract insects with colorful and fragrant flowers to help them reproduce. Larger animals like to eat their fruits and thus spread the seeds of the plants over larger areas. The stationary plants use the ability of others to move around. To achieve this, they use a number of signaling systems, and the most important of them are perhaps the scents.

When a rose is attacked by aphids, the small sap-sucking insects you don't want to see on your precious flowers, it sends out an alarm signal in the form of methyl salicylate. The scent acts as a call to arms to insects feeding on aphids. The signal is broadcast over the garden and gathers ladybugs and other lice eaters who—of course entirely for their benefit—can limit the attack on the rose. Birds can also pick up the scent and get a hint of where they might find nutritious insects to feed on.

The natural world is full of such odorous traces, invisible chemtrails that speak to different organisms with different

Quantum Physics and Life
Ingemar Ernberg, Göran Johansson, Tomas Lindblad, Joar Svanvik, and Göran Wendin
Copyright © 2023 Jenny Stanford Publishing Pte. Ltd.
ISBN 978-981-4968-28-7 (Hardcover), 978-1-003-31267-3 (eBook)
www.jennystanford.com

messages. Insects are the masters of the animal world when it comes to scents. The sense of smell helps them to find food, find partners and to communicate. Their lives are controlled by scent signals. We, humans, are also affected by the scents, perhaps more than we realize.

The sense of smell is the ability to identify chemical substances in the environment. It may be the very oldest of our senses. Fragrance researchers believe that the principles of the sense of smell have been preserved relatively unchanged through millions of years of evolution; the basics of how it works look the same throughout the animal world, from insects to fish and humans. The insects have the ability in their antennas, and we have it in our noses. However, the mechanisms of how a fragrance can become the conscious experience of a smell have been, and are still, debated. This controversy also illustrates more general problems in our understanding of how substances activate receptors on cell surfaces.

Fragrances and aromas are among the most intense pleasures of life. But smells can also cause us to turn away in disgust or suddenly recall memories of childhood. The sense of smell is closely linked to who we are, where we belong and to our survival. Studies have shown that we can recognize close relatives by smell, and some of the most unpleasant smells we can experience are linked to poor food, illness, and death. Fragrances are at work even when we are not aware of them.

But at the same time, the sense of smell carries unsolved riddles. It is a sense with several strange characteristics. One of them is its extremely high sensitivity. For some substances, especially those that are toxic to us, it may be enough for a single molecule to find its way into the olfactory center to trigger a signaling cascade to the brain. Some strongly smelling substances consist of very small molecules. For example, those who played with electric model trains as children have felt the slightly metallic scent of ozone from the sparks that the locomotive created. Ozone consists of three oxygen atoms and is just such a toxic reactive molecule that we have high sensitivity to. Ordinary oxygen gas—necessary for survival—is made up of only two oxygen atoms and has no smell at all.

Nor can we register any smell from the other gases that are constantly around us in the air. Not from nitrogen, not from carbon dioxide or argon. And why should we? It is a change that is interesting, not what is there all the time. Precisely for this reason, blood-sucking insects and ticks are extremely sensitive to variations in the content of carbon dioxide in the surrounding air. Carbon dioxide signals that there is a warm-blooded animal that is breathing nearby, promising a good meal.

The sense of smell is also amazing in its ability to discriminate. It is usually said that humans can distinguish between roughly ten thousand different smells. Some researchers estimate that the number is considerably larger, but we do not know exactly. The sense of smell is also linked to lasting memories. We are bad at describing the smells, but we recognize them, even after decades. And all this even though we have little more than three hundred different receptors in our odor center. So how does it really work?

The body's cells communicate with their surroundings through receptors, molecules that sit on the surface of the cells. There are several thousand variants of such receptors. The receptors of the olfactory system belong to a large family of cell membrane proteins, and they are located in the membranes of the odor cells. These receptors consist of a protruding part that binds to a fragrant substance, a gaseous molecule in the air, and an inner part that extends into the cell. It looks a bit like an antenna that can receive signals from outside.

A fragrant molecule will spontaneously bind to a specific receptor. The physical form itself and the electrical charges in the atoms of the fragrance molecule allows it to bind to the protruding "antenna." It is a bit like a key that fits into a lock where the scent molecule is the key, and the receptor is the lock. The question is how this can lead to chemical activity inside the cell, and then give rise to a nerve signal. How does the inner part of the receptor know that something is happening to the part that is on the outside? It is not easy to see what happens on the molecular level when the scents turn on and off receptors in the olfactory cells. These processes take place in a world that has hitherto been hidden from sight.

To open a lock, it is not enough that the key fits, it must also be turned. So, what is it that turns the key so that the chemical reaction on the outside of the cell causes a signal to be triggered on the inside? The traditional image is that something happens to the receptor molecule. The structure, its shape itself, changes so that it triggers a signal to the inside the cell. In a few cases, we understand how this works, but in most cases, we do not. And this is where the views of the innermost secrets of the sense of smell diverge.

Already in 1928, the British chemist Malcolm Dyson suggested that the sense of smell could be based on the nose acting as a sensor for molecular vibrations and not on changes in the shape and structure of the recipient molecules. Dyson was inspired by a discovery that was brand new in the 1920s, namely that the properties of molecules could be analyzed using their frequency spectrum in a spectroscope. There, light is passed through a sample of a substance, and the light beam produces a pattern of lines revealing which molecules are in the sample. The lines correspond to the quantum mechanical vibrations in the chemical bonds of the molecule, and with such a spectroscope it was possible to identify chemical substances.

Dyson began to measure vibrational frequencies of known fragrances and seemed to be able to conclude that we can actually detect odors by determining the vibrational spectra of the fragrant molecules. Since the structure and shape of a substance did not work very well in predicting how it smells, the nose would instead function as a spectroscope, was the idea. But Dyson's theory was forgotten because there was no conceivable idea of how such a "spectroscope in the nose" could work. It also turned out that some of Dyson's measured vibrations in fragrances were incorrect, which didn't exactly make his hypothesis stronger.

In 1968, two Americans named Lamb and Raklevic discovered that it was in fact possible to make a biochemical vibration sensor, based on quantum mechanical tunneling. The two scientists worked in the laboratories of the Ford Motor Company in Dearborn, Michigan in the days when big car manufacturers spent resources on pure basic research. By studying how electrons could jump between two metal surfaces through tunneling, Lamb and Raklevic discovered that the tunneling current—the flow of electrons—was affected by what type of solvent they placed between the metal plates. They then realized that the spectral features corresponded to the molecular vibrations of the solvents. In other words, they had discovered a way to identify substances with the help of tunneling electrons. Could a molecular "spectroscope" in the olfactory center also be based on this phenomenon?

Luca Turin, a Lebanese-born biophysicist who worked at both the University College of London and at MIT in Cambridge, USA, is a man who is passionately interested in scents and perfumes, but at the same time a driven and bold researcher. In 1996, Turin published a paper presenting a vibrational theory of olfaction. And in 2004, Turin's model was dismissed fairly brutally in the influential journal *Nature* as "a theory that, although provocative, has hardly gained any credibility in scientific circles." But already two years later, the same magazine on its news pages was able to present a study that supported Turin's theory: "A controversial theory of how we feel smells, and which claims that our sense of smell is due to quantum mechanical phenomena, has been thumbed up by a team of physicists."

Since then, Luca Turin has been the leading advocate for a revived image of a sense of smell that depends on how fragrant substances vibrate rather than their molecular shape. In 2011, Turin showed that a substance changes odor as the vibrations in its molecule change. Acetophenone is a fairly common substance used both as a flavoring agent and as an ingredient in perfumes. The smell can be described as pleasantly fruity; some compare the scent with oranges, others with cherries. In an experiment, Turin replaced the eight hydrogen atoms in the acetophenone molecule with heavy hydrogen, deuterium. It is a form of hydrogen with an extra neutron in the nucleus. This does not change the shape of the molecule or any other chemical properties. But the molecule *vibrates differently*, more slowly. This change made fruit flies react differently to the substance. The flies could not only distinguish between the two variants—they could also be trained to avoid one variant or the other. (Yes, you can actually train flies to do simple things, but we won't get into that here.) The flies that were trained to avoid the heavy hydrogen molecule also showed dislike for a completely different molecule, but one that vibrated with a similar frequency. These results supported the hypothesis of a molecular vibration sensing component in the sense of smell—a "spectroscope in the nose."

In 2013, Turin tested whether humans too could distinguish between odorous substances containing common and heavy hydrogen, deuterium. On the one hand, the tests showed that the subjects could not distinguish the two types of acetophenone. But on the other hand, the subjects could easily identify the two variants of other, musky-smelling substances. Turin's explanation is a combination of structure and quantum mechanical vibration. It looks something like this: a fragrance molecule must first match reasonably well with the receptor in the odor center. But in addition, it needs a vibrational energy that allows electrons to tunnel through the receptor and trigger a signal inside the receiving cell. If so, our noses contain a kind of nano-sized spectroscope, that works according to the principles discovered at Ford Motors in 1968.

Turin has also started a company for selling fragrances to the perfume industry—fragrances that can be designed with a computer by analyzing the vibration patterns of the substances. But he gladly admits that his model is not corroborated.

The sense of smell is not only extremely sensitive—it is also very unpredictable. The worldwide chemical industry that makes its profits by developing expensive perfumes or scents for detergents and shampoos has extensive knowledge of the chemical composition of aromas and fragrances. But the chemistry of the scents is very difficult to calculate in advance. It has been virtually impossible to synthesize a new chemical and know how it will smell. Even though companies have gathered accurate data on the molecular structure of more than one hundred thousand fragrant substances, the odors could not be predicted. Until perhaps now.

Using machine learning to draw conclusions from large databases connecting odors and fragrances to molecular structures and shapes, it is becoming possible to say with reasonable accuracy how humans will perceive odors from chemical features of odor molecules. In 2019, researchers at Google published an article describing how they successfully applied AI and deep neural networks to the huge database that scientists previously had not been able to make any sense of.

An important conclusion from those studies is that there is no clear relation between the shape of a molecule and its characteristic smell. Molecules with similar structures could smell very differently, or even be odorless. And molecules with very different shapes could have similar smells. This supports the common knowledge that it is not only the changes in the structure of the receptor that create our olfactory experiences—the old key-in-lock paradigm. That alone is not enough to explain all the peculiarities of the sense of smell. An obvious problem is how to explain why substances that are chemically alike can have completely different scents. A well-known example is mustard and anise—scents that are easy to distinguish with the nose but are triggered by almost identical molecules. At the same time, substances that look completely different can smell almost alike.

The fact that it now seems to be possible to train deep artificial neural networks by machine learning to predict the odor of a given molecule is one thing. But that does not explain how the receptors register a specific molecule, and how the physical olfactory neural network processes the information to produce the sensation of smell. The human nose has millions of olfactory sensory neurons with cell membranes decorated with a number of receptors selected from the available 300–400 different types of receptors. Moreover, each one of these receptors is sensitive to a variety of odorants. Importantly each olfactory neuron only carries one type of these receptors, so each of the hundreds of receptors are distributed on a large number of olfactory neurons. It implies that once one type of receptor is activated, that specific neuron can send an unambiguous "trigger" signal to the brain. The brain thus receives input from a combination of odor receptors on the nerve cells, and the resulting experience from nerve signals in the olfactory bulb to the brain depends on the specific type of trigger at the receptor level. This way of detecting smells makes it possible to distinguish a large number of different combinations, different odors, with a limited number of receptors. One investigator concluded that the brain "interprets smell like the notes of a song." The sequence in which clusters of olfactory neurons switch on decides whether you smell an apple instead of a pear.

When an odorous molecule hits a receptor in the nose, it does not seem to work like a key in a lock. We need to replace that image with a credit card-in-slot model, adding the possibility that a given receptor can distinguish between, but not necessarily identify, several different molecules. At least in the case of insects, recent research results indeed demonstrate that the key-in-lock model is much too simple to be able to describe what is really going on. A cryo-electron microscopy investigation of the jumping bristletail, *Machilis hrabei*, a very interesting insect from an evolutionary perspective, shows that the odor receptor is extremely flexible, and can recognize a wide range of molecules with different structures and shapes. Although there is no obvious need to assume that molecular vibrations play an important role in this organism, the meeting and "handshake" between the odorant and receptor itself does not directly explain the activation by the receptor.

An unsolved problem, which extends to cellular receptors in general, is when there are both agonist and antagonist ligands/molecules that can bind to the receptor. An antagonist is a molecule that can "turn off" or block the receptor, the opposite action to the agonist which "turns on" the receptor. Antagonists are substances that block a chemical signaling system in the body. They can connect to a cell and prevent biological signaling by "getting in the way," as it were, of the usual signaling. For example, there are antagonists to adrenalin; these are the common drugs that go by the name of beta blockers. They prevent the adrenalin from increasing heart activity and blood pressure. Naloxone is another antagonist that blocks the effects of morphine or heroin and can counteract the effects of a dangerous overdose of a narcotic substance.

A theoretical fragrance antagonist would similarly block an odorant before a smell sensation could even start. Such a substance could serve as protection for people working in smelly environments. But so far, all attempts to find a fragrance antagonist have failed. And, to generalize, we cannot yet explain why two similarly shaped molecules fitting in the receptor, give rise to opposite signals, one activating and one blocking the function. Like two similar keys that both fit into the same lock, but only one can activate the lock by turning around.

But then, finally, what happened to Luca Turin's vibrational theory of olfaction? Some would say that the jury is still out, and still others don't expect the jury ever to come back—the jury is hung, deadlocked. It is quite clear that quantum physics is central for describing the close contact between an odor molecule and its receptor. The question is: what is the quantum-related nature of the signal created in the docking process? It might involve a combination of energy transfers involving electron tunneling, molecular vibrations, molecular bending, and changes of geometry.

The story of understanding how fragrances trigger perception of smell illustrates a grey area of cutting-edge science, a problem still in search of a more accurate explanation. To design drugs that will activate or block receptors or create other protein–protein interactions inside cells, has turned out to be a challenge despite solid work. Nevertheless, the rules of the game are slowly emerging. Soon we may be able to understand how our noses can identify a vast world of fragrances, smells and odors. This may eventually shed light on the general problems of how neural signals are triggered or blocked, and how we perceive sensations.

Chapter 13

DNA Repair: Enzymes for Survival and Development

How come we all do not die of cancer already at the age of seven.

—George Klein, Swedish cancer researcher and author

In 1953, the journal *Nature* published what would become one of the most famous scientific papers of the post-war period. In volume 171 of *Nature*, the young British research partners Francis Crick and James Watson proposed a model for the molecular structure of deoxyribonucleic acid. Their model was the now well-known double helix that carries the genes, our inheritance, and the molecule was the one that most people today know as DNA. Before Watson and Crick made their discovery, they knew that DNA was somehow carrying the mechanisms of inheritance, but they didn't know how it worked. The article in *Nature* was the start of a scientific revolution that is still ongoing.

Today we know so much about DNA that we can use sections of the molecule as a unique individual fingerprint. We can read and analyze the complete genomes of extinct animals like mammoths, and with modern CRISPR technology it is possible to replace parts of a selected gene in a cell with high precision. And that is just a small part of it.

Quantum Physics and Life
Ingemar Ernberg, Göran Johansson, Tomas Lindblad, Joar Svanvik, and Göran Wendin
Copyright © 2023 Jenny Stanford Publishing Pte. Ltd.
ISBN 978-981-4968-28-7 (Hardcover), 978-1-003-31267-3 (eBook)
www.jennystanford.com

The long DNA helix resembles a twisted ladder made up of four types of building blocks called nucleotides: adenine, guanine, cytosine, and thymine. They are usually denoted by their initial letters: A, G, C, and T, and it is the order in which these letters appear along the DNA spiral that constitutes "the genetic code." We can now read the entire code of a single individual, with over three billion individual base pairs, in the form A : T or C : G—the steps of the spiral-twisted molecular ladder. The code consists of words made up of three nucleotides, and this is in turn a kind of recipe for the proteins that each individual cell in an organism need to make. A change in the recipe, a mutation in the codeword, can cause the proteins to become unusable or even harmful.

In the introduction we mentioned the Austrian physicist Erwin Schrödinger, one of the true pioneers of quantum mechanics. He wrote in 1940, long before the genetic code was discovered, that "the genetic heritage of our parents is too perfect to be governed by the laws of classical mechanics and quantum mechanics must play a role there." Both Crick, Watson and several other contemporary researchers have testified about how influenced they were by Schrödinger's thoughts. The structure and function of the genes became the big mystery that begged to be solved.

Today we know quite accurately how the mechanisms of inheritance work, how the long lines of genes A, G, C, and T are translated into messenger molecules that collect amino acids in the cell and how these then are assembled into new proteins in the protein factory units called ribosomes. Sometimes the assembly goes wrong if the sequence of nucleotides changes. This can be caused by radiation, free radicals, or toxic substances. *But most of the time the errors happen spontaneously.*

This model for spontaneous mutations was suggested already by Crick and Watson, and further developed by the Swedish chemist Per-Olov Löwdin in the 1960's. It was Löwdin who showed how the transition was possible if protons were tunneling into places where they weren't expected to turn up. (It was also this process that prompted Löwdin to use the term "quantum biology," a concept that was coined already in the 1920s, but now for the first time applied to a specific biomolecular mechanism.) The idea of spontaneous displacement and the mutations that followed wasn't confirmed until quite recently by using techniques that can capture and visualize these fleeting and short-lived activities on the atomic level. Many molecules change their structure depending on the interaction with their immediate environment. Like the retinal molecule in our eyes, which bends after being hit with a photon and triggers a nerve signal to the brain, as described in Chapter 9. These different molecular shapes are called isomers; they have the same chemical composition but a different structure, and that difference can influence the molecular surroundings.

The nucleotides of the DNA ladder can also become isomers—slightly changing shapes. They do it rarely, and during extremely short time intervals, but if this happens during the copying process it can result in a typo, a change in the sequence of letters. The temporarily altered structure creates a bond with the wrong base, an A with a C or a G with a T. The change that brings about such a mutation lasts for less than a millisecond, and then the nucleotide falls back to its normal shape. This model that was suggested in the 1950's wasn't confirmed until quite recently by using techniques that can capture and visualize these fleeting activities on the atomic level.

The individual base pairs of DNA are held together by hydrogen bonds, a connection where an electrically positive hydrogen ion attracts a negative partner. This kind of bond connects the steps of the DNA ladder, and it holds together because electrons and protons tunnel in the bond between the nucleotides. During short moments this tunneling can take place with the "wrong" partner, because of the altered shape of one of the nucleotides. This means that the spontaneous mutations are the result of transitions on the quantum level, random fluctuations in the atomic world.

But if the information-carrying DNA molecule is so exposed and so sensitive, how can our genetic material be preserved? How can the information system work with a hard drive that is constantly damaged? All these mutations should break down DNA and cause serious diseases in the long run. Mutations, errors in the DNA string are a basic mechanism behind cancer development. The Swedish cancer researcher and author Georg Klein once asked the question: "how come we're not all dead of cancer at the age of seven?"

The explanation, and thus our salvation from an early death, is that there are efficient repair systems that constantly restore the code—systems that provide constant error correction to secure the information in the genes. Our knowledge about genes and heredity has exploded since the 1950s. One of the more important discoveries that made much of today's DNA technology possible was made in the 1980s by American Kary Mullis. He is in many ways an unusual scientist. Among other things, Mullis claims to have met an alien in the form of a raccoon, and he believes that astrology is a source of real knowledge. But what makes him most interesting in this context, though, is his research.

Thermus aquaticus is a bacterium that lives in hot water. It was first discovered in one of the geysers of the Yellowstone National Park in the United States. This heat-loving bacterium carries a substance capable of multiplying sequences of the genetic material, an enzyme called polymerase. This enzyme is a molecular machine that quickly creates thousands of copies of single DNA pieces. It has proven to be an indispensable tool for DNA analysis, for disease diagnoses and crime scene investigations, as well as for mapping the genes of animals and plants, or prehistoric people such as the Neandertals. This was realized by Kary Mullis, who developed a technique for using polymerase during the 1980s. And for that he received the Nobel Prize in chemistry in 1993.

The polymerase enzyme can transform small amounts of DNA into useful quantities by copying and re-copying the sequences of interest. All cells contain polymerase, as it is a natural part of the cell's life to divide and copy its own DNA. The special feature of the Thermus bacterium's polymerase is that it is heat resistant, which is crucial for the method to work at high speed. And it acts fast. At 70 degrees Celsius, the bacterial polymerase can copy one thousand base pairs in ten seconds. It thus assembles one thousand of the A, C, T, and G which are the letters of the genetic code and which are the building blocks of the genes of all living creatures from bacteria to blue whales, ants, and humans.

Polymerase is one of Life's many active molecules, specially designed by evolution, and fabricated inside living systems. Furthermore, the copying enzyme of the Thermus bacterium is just one out of a whole universe of enzymes that drives the nucleic acid copying processes in living cells.

Enzymes are almost everywhere. Anyone who lacks the enzyme lactase cannot make use of lactose—and thus becomes lactose intolerant. Amylase is an enzyme that breaks down starch and that starts to work already in the oral cavity when we eat vegetables and fruits. Without the enzymes that hide in yeast, we wouldn't have bread—or beer and wine for that matter. And by far the most common enzyme on Earth is one that goes by the name rubisco which helps the green plants to convert sunlight into carbohydrates in photosynthesis.

Enzymes are catalysts. They make things happen without being consumed themselves. The vital reactions that happen within milliseconds in a cell would often take years to happen without the help of the enzymes. In fact, enzymes being the chemical workhorses in all cells are at the core of what is specific for life; the ability to drive useful reactions far from equilibrium. Enzymes ensure that cells can respond to external signals, that muscles can be activated, and that our food turns into energy. They make sure that the respiration we describe in Chapter 11 works smoothly at 37 degrees Celsius.

In the cell's organized chaos of molecules, the enzymes act as large molecular assistants. They speed up events and they do it efficiently. Chemical reactions often require energy to get started. This activation energy is a threshold that must be overcome for a reaction to start. A push over the threshold is needed. Paper spontaneously burns in air, but the reaction—fortunately—does not start until you put a flame to the paper. The heat energy from the flame activates the reaction which then continues on its own until the paper has burned up. A catalyst, such as one of our enzymes, makes the activation threshold lower. It, therefore, has the useful ability to trigger reactions that wouldn't occur spontaneously.

The enzymes have evolved through hundreds of millions of years of evolution. Each small improvement provides a survival benefit and will stay in the gene pool. By utilizing tunneling, an enzyme catalyst can "penetrate" activation barriers and achieve maximum speed in important chemical reactions. Quantum processes are thus an essential ingredient for many vital processes to take place at all. But they are also as we have seen, indispensable for life's ability to constantly change and adapt.

But if the quantum processes constantly degrade DNA, how do we answer Georg Klein's question, why don't we get cancer already in childhood? It was this mystery that fascinated the Swedish biochemist Tomas Lindahl as early as the 1960s. Tomas Lindahl discovered that both RNA and DNA were extremely sensitive molecules when he worked with them in the laboratory. It was difficult to get the molecules to keep their structure intact when he examined them. But these were the two substances that form the basis of life. They ought to be extremely stable, but they were not. The DNA molecule is so unstable that life on Earth would be threatened if it isn't somehow protected. And, fortunately, it is.

Our rescue from an early death is the effective repair system that constantly restores the code. Systems that are constantly proofreading the genes after their sequences have been copied. Evolution requires that mutations occur. Without changes in the genetic information, organisms cannot adapt to the environment. But we can only tolerate some single genetic news in each generation, so that evolution can select the changes that are beneficial for survival and reproduction—the process Darwin called natural selection. Larger mutations, or too many changes in the genetic code, usually lead to serious problems—illness and death. And it is the repair mechanisms—the error correction—that keep the number of disorders in the DNA sequence at a manageable level.

Three important molecular repairing processes keep our DNA in stable condition. The first is called *base excision repair*—repair by cutting off bases. How it works was mapped by Tomas Lindahl in bacteria in the 1970s. Lindahl found an enzyme that removed the damaged part of cytosine—the letter C—and that eliminated the mutation. Later, he has also shown exactly how this process goes on at the molecular level even in human cells.

A common cause of DNA damage is ultraviolet radiation from the sun. Ultraviolet light is short-wave with high energy and can affect the chemical bonds in the DNA structure. The pigment melanin is a protection mechanism that darkens the skin and gives a suntan. But too much sunlight can cause serious damage. Life arose in the oceans where the sun's rays were filtered off by the water. About 350 million years ago, the first animals ventured up on land at least for shorter periods. In order to take that step, they needed organs to breathe air, and they also needed to protect themselves from the constant ultraviolet radiation.

The second type of repair mechanism that evolved is called *nucleotide excision repair*—it cuts out chemically modified nucleotides. Bacteria have two systems to do this. One of them works with light, and one works in the dark. Three different enzymes in the cell can find and identify damage caused by ultraviolet light, and also make cuts in the DNA strand—one on either side of the lesion—and remove the damaged piece. This repair system employs quantum tunneling of electrons. By utilizing tunneling, an enzyme can achieve maximum speed in important chemical reactions. The enzymes that control the repair of UV damage work on time scales of picoseconds, a trillionth of a second, 10^{-12} seconds. When the enzymes absorb photons from blue light, it cuts the DNA chain so that the damaged part can be eliminated.

The third form of DNA repair is called *mismatch repair*—repair of mismatches in DNA base pairing. If one DNA strand contains a C and is paired with a T, base pairing does not work well, and a small bubble occurs in the DNA ladder which is recognized by a particular enzyme that causes base pairing to be corrected. Of the thousands of errors that occur when the human genome is copied, mismatch repair rescues most people.

In addition to these three mechanisms, there are several others that do daily repair to the DNA damage caused by the sun, by cigarette smoke or other harmful substances. At each cell division, mismatch repair corrects a thousand matches that have happened to be wrong. The 2015 Nobel Prize in chemistry was shared between Tomas Lindahl, American Paul Modrich and the Turkish-born Aziz Sancar for their work on exploring and describing these different ways for the cell to repair DNA.

Without the repair system, we would get one mutation per million base pair during each cell division, that is several million mutations in each cell. The repair system reduces the mutation rate one million-fold to one mutation per trillion base pairs in every division, resulting in *one mutation during each cell division*, a level the cell can handle. Over the years, we collect inaccurate codes in the genes that usually do not cause disease, but contribute to aging. Lack of DNA repair can lead to a disease called Werner's syndrome which causes premature aging with atherosclerosis and cancer. Many forms of cancer are caused by some of the repair systems being partially inactivated. It also means that the DNA of the cancer cells becomes unstable and easily changes. That is why cancer cells often mutate and quickly develop new properties, such as resistance to chemotherapy. But with a defective repair system, the cancer cells are also extra sensitive. Completely without repair mechanisms, their DNA will easily be damaged, and the cell will die. This sensitivity can be used to create anti-cancer drugs. By knocking out the repair systems that are still active in the diseased cells, it is possible to stop the growth of cancer cells. Perhaps the principles of quantum physics in the future can also be used to fight cancer, as discussed in the next chapter.

Chapter 14

Quantum Physics in Diagnosis and Treatment

Logic will take you from A to B. Imagination will take you everywhere.

—Albert Einstein, Nobel laureate in physics, 1921

One of the best-selling books in London in 1665 was a description of a world invisible to the eye. Robert Hooke had built a microscope and depicted in his work "Micrographia" what he saw through the lenses. Hook's drawings of a magnified flea turned a mite into a monster. He showed in detail what the foot of a fly looks like, and saw that plants were made up of lots of small rooms that he called cells. These were pictures that could both frighten and arouse interest among the educated bourgeoisie that read the book. What more might be hiding in the unknown little world? Well, it would take nearly another 400 years before we could see how single molecules in the cell carry on the game that is life.

Light consists of photons. But at the same time, as we know, it is a wave motion. This means that there is a natural limit to how small things can be and still be visible to the naked eye, even under a microscope. Visible light has wavelengths of around half a micrometer. A biological cell is typically ten micrometers in diameter. In other words, it is about 50 times bigger, much like comparing a tennis ball to a car. When considering such relatively large objects

Quantum Physics and Life
Ingemar Ernberg, Göran Johansson, Tomas Lindblad, Joar Svanvik, and Göran Wendin
Copyright © 2023 Jenny Stanford Publishing Pte. Ltd.
ISBN 978-981-4968-28-7 (Hardcover), 978-1-003-31267-3 (eBook)
www.jennystanford.com

as a cell, one can treat the photons as classical particles bouncing in straight lines according to Newton's optical laws. But if you want to see something really small, of a size much smaller than the wavelengths of light, then the photons will behave as waves. They will simply move around the object without changing direction, and the object will be invisible. Consequently, if you look at an object that is less than a few hundred nanometers in size, it will look like a blurred spot of light. Somewhere around a few hundred nanometers is the limit for how small things can be seen even with the very best traditional light microscopes.

One way to get past this limit is to use an electron microscope. It uses electrons instead of photons to illuminate what you want to study. And quantum physics has shown that although electrons are particles with mass, they simultaneously behave as waves. All particles can be associated with a de Broglie wavelength that is inversely proportional to the mass and speed of the particle. This means that the wavelength becomes shorter the higher the particle speed. Very fast electrons have a much shorter wavelength than visible light, which means that they will bounce off much smaller objects, not just passing around them like waves. They can therefore help us see much smaller things than biological cells—even the atoms and molecules inside the cells.

The Nobel Prize in chemistry 2017 went to three researchers who invented a method that allows you to see the structure of individual protein molecules when the tissue sample is frozen solid—so-called cryo-electron microscopy. In practice, the cryo-electron microscope works as a high-speed camera that captures events by freezing molecular motions that occur during picoseconds. By looking at many frozen images one can then get a picture of the dynamics, really see how the molecules move. It is one of the most powerful methods available today to study the structure of large biological molecules and systems—proteins, viruses, bacteria—at the atomic level. The beautiful recent pictures of the nasty, currently active, Zika, Ebola, and Corona viruses are produced with cryo-electron microscopy.

With femtosecond laser light flashes, the motion of atoms and molecules can be frozen while forming chemical complexes, such as biomolecules. This means that for the first time we can "see" the short-lived super-molecules that exist for brief moments during the reaction, and which then develop into their final products. In this way, it has been possible to study the extremely rapid processes that cause visual impressions in the retina, from Chapter 9, or those that drive the photosynthesis, from Chapter 10.

Another way to penetrate Life's hidden secrets is to use the light-emitting green fluorescent protein (GFP). This protein was originally derived from a small harmless jellyfish that lives in the waters along the North American Pacific coast. The jellyfish can emit a green glow to attract prey. With the help of the luminous protein, it is possible to study cells or processes within cells. GFP is a tool that makes it possible to follow the life processes of cells with the naked eye. The ability of the protein to shine, fluoresce, is itself a quantum physical phenomenon. It occurs when photons of a certain wavelength hit electrons in the protein and lift them up to a higher energy level. When the electrons fall back, they re-emit photons, usually with a red-shifted color because the electrons lost some energy to molecular vibrations before falling back.

Anyone who has put a tube to the mouth and sung a note knows what resonance is. If you sing the right note, the sound gets stronger. You hit the tube's resonant frequency and the air in the tube begins to vibrate with larger amplitude. And when the frequency of a radio transmission matches the resonant frequency in the radio receiver circuit, the electronic circuit begins to oscillate in tune and the music flows out of the speaker. The conditions that apply to the tube, the sound waves, and the radio receivers also apply to how atoms and electrons relate to electromagnetic radiation. Everything from radio waves and microwaves to visible light and X-rays can create resonances in the particles. When the frequency of the radiation corresponds to one of the atom's resonant frequencies, parts of the electron cloud begin to oscillate more strongly. This is the phenomenon that is the basis for the so-called magnetic resonance cameras, based on nuclear spin and nuclear magnetic resonance (NMR).

Magnetic resonance imaging (MRI) is based on quantum physics. Protons, neutrons, electrons, and many other elementary particles behave like small magnetic compass needles that are free to point in different directions, and which can orient themselves along a magnetic field, just as a compass needle points in a north-south direction along the Earth's magnetic field. The magnetism of the particles comes from the quantum physical property called spin, which we discussed in Chapter 4. Although the concept of "spin" makes it sound as if the particle is spinning, it has nothing to do with rotation in the classical sense. The spin is a purely quantum mechanical phenomenon that must be described with quantum physics.

The simplest atom of them all—the hydrogen atom—has only one proton and one electron, both of which have their own spin. Larger atoms have several protons, neutrons and electrons, and the atom's spin is the sum of the total nuclear spin and the total electron spin. Thus, in a molecule with many atoms, we can say that there are magnetic compass needles at the positions of the atoms, needles that can point in random directions. But by applying a magnetic field to the molecule, you can align the compass needles in the same direction. They adjust according to the direction of the magnetic field. And applying electromagnetic radio or microwave radiation, you can make the compass needles oscillate in tune with the frequency of the radiation. When the frequency of the radiation hits the resonant frequency of the atoms, the atomic nuclei become excited. The nuclear spin will then flip up and down and absorb, transmit and disperse microwave radiation. When you get a medical examination with an NMR machine, the powerful magnetic field aligns the hydrogen atoms in the body's water and adipose fatty tissue. When the resonance occurs, the tissues send back radiation that can be used to construct detailed images of the body's interior.

Nanos is Greek for the word "dwarf," and nanotechnology is a collective name for many practical applications where extremely small components are used—so small that the objects you work with can almost be compared to elementary particles. And particles, as we have seen, can behave according to the principles of quantum physics. Nanoparticles can now be found in many products, from bicycle frames and tennis rackets to makeup. They make the materials lighter and stronger, make surfaces repel dirt, make food durable, or kill bacteria in textiles. But the properties of the nanoparticles can also be utilized for scientifically more advanced purposes. In medicine, there is intensive research toward two goals: to use the quantum mechanical properties of the nanoparticles to make accurate diagnoses and to fight deadly cancer cells.

Cancer is caused by a technical defect in one of the genes in the cell's DNA. The longer we live, the greater the likelihood that any of our billions of cells will end up with a harmful mutation and start running amok. In the worst case, such a mutation causes the cell to start producing far too much of a protein or create proteins that are completely mis-engineered. For a long time, there was nothing one could do with cancer other than try to cut the tumor out of the body whenever possible. Often this was in vain, as the cancer had already spread to other parts of the body.

About 100 years ago, the German medical researcher Paul Ehrlich discovered that certain substances could stop infections by killing the bacteria that caused them. Germany's Emperor at the time, Wilhelm II was a known hypochondriac, and he is said to have asked Ehrlich if his method could cure cancer. It could not, and the crucial problem, Ehrlich realized, was that his toxins could find and kill bacteria just because the bacteria were so different from human cells. Cancer cells, on the other hand, are almost identical to healthy human cells. It proved difficult to find a substance that would only attack a tumor, without the poison affecting the entire body.

This is a concern that has stayed with us despite the development of better and better chemotherapy to fight tumors. The cytotoxins used often cause severe side effects. The most obvious one is that the treated patients often lose their hair. But more serious effects are, for example, that the immune system is exposed and compromised, and the person being treated must therefore be isolated and protected against infections that could otherwise be fatal. If instead, the cell toxins could be concentrated only to the exact places in the body where they are really needed, and leave the other cells alone, the treatment could be both more effective and gentler. This is where the nanoparticles come into play. The goal is to create particles that turn into target-seeking missiles for attacks against harmful cancer cells and the delivery of medicines with high precision.

The nanoparticles are very small, and it is by utilizing the quantum physical principles of the nanoworld that the particles can become effective weapons in the fight against cancer, in more ways than one. It has been found that nanoparticles tend to accumulate precisely in tumor tissues. This is because the fast-dividing lumps of cells that make up a tumor have a different architecture compared to healthy organs. All tissues in the body need oxygen and nutrition that is delivered by the blood, and this is just as true for cancer cells. But the fast-growing tumor is unable to build the fine network of thin blood vessels that do the job in a healthy body. Instead, it becomes a jerry-building, a leaking jumble of vessels built with no consideration of the future. And in this poorly drained environment, nanoparticles will accumulate if they are injected into the body.

Here, too, one may work with resonance oscillations. The nanoparticle contains an electron cloud that can be affected by electricity. If you drive the particle with an electric field that constantly alternates in polarity, the electron cloud of the particle will oscillate together with the field. But when the frequency of the field approaches a certain value something new happens—the nanoparticle comes into resonance and the amplitude of the oscillations increases sharply. This allows the nanoparticles to be externally driven to treat a cancerous tumor in several ways.

Using electric fields of variable frequency it is possible to set the particles in oscillation with resonance frequencies that can be determined in advance by adjusting the sizes of the particles. Metal nanoparticles can be made to oscillate employing resonant radiation so that they heat up and kill the tumor. With semiconductor nanoparticles, one instead gets a response in the form of fluorescence—the tumor begins to glow, revealing exactly where it is located and how large it is.

Once in place, the nanoparticles can not only absorb, but also emit and reflect light. Moreover, they can supply medicines in the form of cytostatic drugs. The cell-killing substances may be attached to the surface of the particle or lie embedded inside a "nanosphere."

Another method is to use magnetic nanoparticles sent through the membrane of the cancer cell. The cell wall then yields and encloses the nanoparticles, forming a small artificial cell. The nanoparticles are thus encapsulated in one of the cell's natural transport containers—the lysosome. Once inside the cancer cell, the nanoparticles can be rotated using a rotating external magnetic field—like in an induction cooktop—to tear the fragile membrane of the lysosome to pieces and spread cell fragments and nanoparticles. This then causes the cancer cells to die through the process called programmed cell death, or apoptosis.

All this is promising research for the future—the use of nanoparticles for diagnosis and therapy is still at a pioneering stage.

Experiments at the forefront have mainly been performed on mice and on human cells in a petri dish. Extensive studies are required in each case to apply it to humans. Living cells have a fantastic natural ability to both incorporate and exclude particles and molecules of various kinds, and it is difficult to predict how this type of nanoparticle will be handled by such a complex living biological systems as humans.

But, most importantly—nanoparticles do not themselves provide a cure for cancer. Their role is to bring genetic and other tools of molecular biology to the locations of the cancer cells for direct preventive action on the metabolic processes that allow the cancer cells to grow and spread.

Chapter 15

No More Mysterious Than Necessary

To see a World in a Grain of Sand
And a Heaven in a Wild Flower,
Hold Infinity in the palm of your hand
And Eternity in an hour.

—William Blake, *Auguries of Innocence*

Popular scientific descriptions often emphasize concepts and phenomena that are difficult or "impossible" to understand and sometimes make them sound more mysterious than necessary. We want to avoid that. In this book, we have talked about how quantum physics appears in concrete contexts at the atomic scale, for example in living cells. We have also discussed how quantum physics must have played an important role early in Evolution, perhaps even when life arose. But will quantum physics also influence the higher life processes? Can it affect how we perceive the world around us, and can it affect our free will?

The classical and quantum mechanical worlds work in different ways, but where is the boundary between the two worlds? Can quantum-mechanical phenomena such as tunneling and entanglement emerge in visible everyday life? Does it ever happen that the tennis ball that we used as an example in Chapter 4 penetrates the wall and suddenly emerges on the other side? These are questions that have fascinated scientists and philosophers for

Quantum Physics and Life
Ingemar Ernberg, Göran Johansson, Tomas Lindblad, Joar Svanvik, and Göran Wendin
Copyright © 2023 Jenny Stanford Publishing Pte. Ltd.
ISBN 978-981-4968-28-7 (Hardcover), 978-1-003-31267-3 (eBook)
www.jennystanford.com

almost one hundred years. The answer from a physicist is that it is a question of how well a quantum system can be kept isolated from its environment.

If a system is completely isolated in a physical sense, like in a box with very thick walls, it is necessarily quantum mechanical and coherent. However, if the system is open to the environment, sooner or later it will lose its coherence—energy and entropy are disseminated into the outside world and quantum phenomena become blurred.

This is exactly what happens in the stories we have told in the previous chapters, where quantum physics describes what happens in the retina of the eye or the navigation of birds. There, we see how quantum phenomena can occur in the cell's open system if the coherence—the "flat water surface" where quantum mechanical waves can propagate undisturbed—lasts for a sufficiently long time to have a real biological effect.

What is meant by an isolated system in the world of physics cannot really be described in everyday terms. A completely isolated room must have walls so thick that nothing—photons, electrons, protons, or neutrons—can tunnel through. And if a human being were to be trapped in such an imagined isolated box, its existence would not be transformed in any quantum physical way. We would not escape our everyday physical laws: the total isolation would soon lead to an unpleasant death—for lack of oxygen if nothing else.

Just like electrons and photons, large objects like humans, cars, and tennis balls, also have a double nature in principle, as we described in Chapter 4. In other words, they can be both waves and particles. Living creatures—and we ourselves—actually have a wave nature, but our de Broglie wavelength becomes so unbelievably small that it is meaningless in this context. In every practical sense, it is negligible. The only way for a person enclosed in a box to "experience" the world quantum mechanically, is if the person's own wavelength is comparable to the size of that box. This is indeed completely impossible—a man's wavelength is many orders of magnitude smaller than the size of an atom. It would require the room to be the size of a quark, or smaller. And it would then be rather uncomfortable to squeeze inside.

Furthermore, in an isolated system where the laws of quantum physics prevail, time will lose its direction. Everything becomes reversible—that is, all processes can also go in the opposite direction. It does not mean that the trapped man lives forever in his box. Man is a macroscopic system that would soon perish. We can never reverse the direction of time without first turning into particles with quantum properties and coherence time. And that is by no means anything to strive for.

But how about the concept of entanglement? The one we described in Chapter 4, which, for example, may exist in the retina of birds and that might help them and other animals to navigate?

Entanglement may seem to imply instant transfer of information, faster than light and with no cost of energy. Chinese scientists used a satellite in 2017 to link two ground stations and showed that two photons could maintain their quantum mechanical entanglement over a distance of 1 200 kilometers. A new world record at the time. But the result *did not imply* that the photons "communicated instantly" with each other.

The two entangled photons were created on the satellite with a laser in a common quantum mechanical state and sent down to two different receivers, located at Delingha and Lijiang, 1 200 kilometers apart in Tibet. The photons remained entangled in the same quantum mechanical state as on the satellite, despite traveling to widely separated locations. But in order to establish this fact, classical radio communication between Delingha and Lijiang was required. The two stations simply needed to call each other on the phone and ask about the result. Only then could they conclude they had indeed shared an entangled state. No information was transmitted faster than the light. Entanglement thus does not open for instant communication over long distances. And it also does not open for teleportation of the kind we know from, for example, the TV series Star Trek, where people can "beam" between two places at an instant.

In fact, it is possible to teleport small quantum states. But doing it for large quantum states is very difficult, and for macroscopic states—for example, Captain Picard in Star Trek—it is simply impossible. Here one can note that in what is called quantum teleportation, it is only *the quantum state* that is teleported, not the matter itself. One needs to have an identical system at the target station, which for the state of single atoms that have been teleported so far is not a problem. However, to teleport Picard, one would first need to create an atomically identical "golem" version of Picard which is fully entangled with the original Picard. This golem must be transported physically to the target location. The teleportation is then performed when the original Picard is being measured completely and destructively. All these steps are simply impossible.

It is not uncommon to hear that quantum physics shows that the world is governed by chance and uncertainty. We do not know exactly which side of the metal rod an electron will choose when conducting Tonomura's experiment from Chapter 4, and according to Heisenberg's uncertainty principle, it is not possible to know where a particle is at the same time as we measure its velocity. This means that quantum physics can be used as an argument that the world is actually completely unpredictable.

But just the opposite could in fact be true.

In quantum physics—we say it again—a particle is connected to a wave motion, a wave function in the terminology of physics. The wave function is often described as a superposition of quantum mechanical states, that is, the particle-wave is simultaneously in several states. Like hitting several keys on a piano to create a chord. But this description is only meaningful when we are looking to detect and measure some of the different states of the particle. Like filtering out a specific tone from the chord struck on the piano.

It is the properties of the measuring device that offer the many possibilities the particle must adhere to when it is detected. The particle must select one, can only select one of these states when it is detected by the measuring device. This is what is called the "collapse of the wave function" due to the measurement. But that does not mean that the particle collapses as an object. If you prepare a similar particle and repeat the experiment, the same measurement will result in the collapse of the wave function to another of the possible states. If the experiment is repeated many times, you will see a statistical distribution of the results of the measurement. The graph describing the distribution then corresponds to the actual state of the particle before the measurements.

A good example is a famous experiment from 1922. There, a beam of silver atoms was sent into the gap of a specially shaped magnet so that the atoms were bent upward or downward depending on whether the spin of an atom was directed along or opposite to the direction of the magnetic field. When the atoms then hit a photographic plate, two spots appeared on the plate, one upper and one lower.

It is in such situations that quantum mechanics rules and classical physics does not work. When one sends in a single atom where one has prepared a pure spin-up or spin-down state, one always detects a single small point in either the upper or the lower spot area with 100 percent probability. Instead, when the atom is prepared so that the spin is perpendicular to the direction of the magnetic field, the atom will hit either the upper or the lower spot with 50 percent probability. After many attempts, the spots will be the same size and black.

The explanation is "simple" if you make use of quantum mechanics. An atom with perpendicular spin is in a superposition of equal parts of spin-up and spin-down. In principle, each atom remains in this superposition state right up to the plate. Only when the atom hits the screen does the state collapse to spin-up (upper spot) or spin-down (lower spot). Which path the atom chose is not known—in fact, it took all the roads to the screen. In the same way as did the photon in the chlorophyll of plants, or in the cryptochrome in the retina of the eye. The result of the detector is a purely statistical process.

This statistical result is the "uncertainty" in quantum physics. We cannot predict which path the atom should take. One could say that this is a way of formulating what is called the Copenhagen interpretation of quantum physics, with Niels Bohr in the 1920s as the foremost representative; it simply describes how it works—"shut up and calculate." There is no explanation that can be expressed in classical terms using our everyday language. However, many people find this extremely unsatisfactory. Albert Einstein, for example, who once again appears in this story, being one. He refused to accept that God "plays dice with the universe."

The fact that a quantum mechanical system does not "show its colors" until we observe it has had philosophical consequences; one has asked if anything at all exists before we measure and force it into a particular state. If the very existence of a particle depends on our observation of it, then what is real? The question easily leads to the idea that we create reality by describing it, that is, our image of the world is not only a subjective experience, but that we actually affect the world directly with our observation.

But after all, it's hard to avoid that reality exists regardless of what we think of it. The properties of a quantum mechanical system—for example, a water molecule—are described by Schrödinger's differential equation, one of the basic mathematical expressions of quantum physics. The water molecule bounces around in its environment—perhaps inside a biological cell—rotating, vibrating, binding to other molecules. In relation to its surroundings, it is highly real. If we were to have the chance to meet this particular molecule and ask for its state, it must give a definite answer. But just because the water molecule's wave function thereby "collapses" to a certain state does not mean that the water molecule itself collapses—it remains highly real.

Finally—is everything quantum mechanical?

Niels Bohr's view of the world is based on a crucial assumption: that there exists a genuinely classical world parallel to the quantum world. The electrons and tennis balls must be described in different ways—the tennis ball with Newton's mechanics and the electron with Schrödinger's or Heisenberg's quantum mechanics. According to Bohr, the two worlds are basically separate, and it is difficult to convincingly explain—even for the Copenhagen School—how the quantum world transitions into the classical world. Exactly when and how does a collection of quantum mechanical molecules turn into a classic tennis ball?

Another way to illustrate this difficulty is to introduce the perhaps most well-known pet of modern science history: Schrödinger's cat. Erwin Schrödinger, the Austrian physicist who appears again in the story here, in 1935 formulated a thought experiment to shed light on the consequences of using quantum mechanical principles in our everyday world.

In short, the thought experiment looks as follows: A cat is isolated in a box along with a device that may, or may not, trigger a poison vial. The outcome is controlled by a radioactive disintegration process—that is, according to the well-known laws of nature, a completely random event. If the radioactive substance emits radiation, then the mechanism is activated and the poison—hydrogen cyanide—spreads in the box whereupon the cat dies. After waiting for an hour, we do not know what has happened inside the box. The cat may be alive, or already dead. And since we don't know until we look, we can pretend that the cat is in a quantum superposition—both dead and alive at the same time. According to quantum mechanics, it is not until we observe the cat that it will "collapse" into one of the states, dead or alive.

But despite this famous parable, cats in superposition were an absurd thought for Schrödinger, as well as for Einstein. Nevertheless, this cat has become more long-lived than both Schrödinger and Einstein probably imagined. Schrödinger's paradox poses a question that has received many different answers over the years. No matter what happens inside the box, physicists have continued to debate how we should view the cat's condition. Does it really end up in a state of "neither living nor dead"? What happens if we place a human in the cat's place? Or a bacterium? What does it experience? Is it any different if we let a computer record what is happening?

An alternative way of looking at it all is to assume that everything remains quantum mechanical instead. The classical world then becomes a special case of the quantum world where formal quantum mechanical things can be observed, documented, and copied in an unambiguous—that is, classical—way. This is a theory formulated by Wojciech Zurek at Los Alamos National Laboratory in the United States and named quantum Darwinism. The theory is still mainly a significant contribution to the debate, but there are some experimental results that support the idea, and it possibly describes some important aspects of reality.

We can exemplify the consequences of Zurek's theory with Schrödinger's cat. A macroscopic quantum system in the form of a biological cat can reasonably be nothing but living or dead—not in a superposition of living and dead. But a small, isolated quantum system—for example, a small quantum processor with 15–20 quantum bits—can be put in superposition of all possible quantum states. The problem is that if the quantum processor is made much larger, it may be difficult to isolate the quantum system from the environment. The system's quantum bits may quickly get entangled with the environment and the quantum processor's initially unique properties will not be distinguishable from the environment. The fine, delicate superpositions in the quantum processor will quickly die out—what we call decoherence—and there will remain a small number of long-lived states that correspond to the cat being either dead or alive. The quantum processor has in practice become a classical system.

The very concept of quantum Darwinism comes from Zurek's theory that over time, the more persistent and stable quantum mechanical states become reinforced—a kind of "natural selection." This ultimately defines a classical world that we recognize from our daily experience of everything from flowers to geckos. Although it is, according to Zurek, fundamentally quantum physical.

But what does it matter if we theoretically understand how the quantum world transitions into our common world? Are there any practical consequences?

Actually yes. One can follow a quantum mechanical system in time and at the same time disturb it minimally. If one collects enough information about its development, then you can also control the system, counteract the decoherence and even reverse the development. Experiments can even follow a system through a spontaneous change of state—a quantum leap—and then in principle restore the system. This will be crucial for the future development of quantum communications, quantum computers and sensor systems.

Chapter 16

Consciousness: The Greatest Mystery

The great mystery of our consciousness is beyond our grasp.

—William Shatner

I would argue that nothing gives life more purpose than the realization that every moment of consciousness is a precious and fragile gift.

—Steven Pinker

I think the brain is essentially a computer and consciousness is like a computer program. It will cease to run when the computer is turned off. Theoretically, it could be re-created on a neural network, but that would be very difficult, as it would require all one's memories.

—Stephen Hawking

Ego is the immediate dictate of human consciousness.

—Max Planck

Lizards hang in the ceiling and birds and butterflies find homes over great distances, all thanks to quantum physics. The sensitivity of our eyes and the vitality of the green plants are based on the same quantum mechanical phenomenon. We have shown several examples where quantum physics can have decisive effects on sensory impressions and on the molecular machinery of cells. And

Quantum Physics and Life
Ingemar Ernberg, Göran Johansson, Tomas Lindblad, Joar Svanvik, and Göran Wendin
Copyright © 2023 Jenny Stanford Publishing Pte. Ltd.
ISBN 978-981-4968-28-7 (Hardcover), 978-1-003-31267-3 (eBook)
www.jennystanford.com

we have found that quantum physics is likely to come into play when it is useful in the evolutionary game that is life on Earth. Several of these phenomena appear to have been active since the very beginning of life's history. For example, efficient photosynthesis, which is a prerequisite for the entire biosphere we are part of today. That mechanism originated billions of years ago.

But if quantum physics is important for several sensory organs and vital events in the cells, is it not reasonable to expect that it also may affect our cognitive activities? Human consciousness has been described as the last major challenge in research, the really difficult question or the hard problem, to quote the American philosopher David Chalmers. How do we experience the world, have feelings, and see ourselves from the outside? And how does that character we usually call Myself come to be?

Our lives begin as a single fertilized egg cell and a few decades later, after more than 70 cell generations and millions of cell divisions, we end up as a biological system of immense complexity, an organism that can make fire, play games, or discuss Einstein's critical views on quantum physics. When during that journey do we become aware of ourselves?

No scientist or philosopher today can claim to really understand how our consciousness arises out of the brain's networks of nerve cells. We also do not know if it is something uniquely human. There are several examples of animals being able to show signs of consciousness, even if their "intelligence," that is the ability to solve problems or handle unfamiliar situations, is not as sharp as ours. British cognitive scientists Lars Chittka and Catherine Wilson argue that even insects are conscious. Several experiments with bees indicate that they not only react with reflexes or innate patterns of action but can assimilate new and unexpected information and act on it. The two researchers see consciousness as an evolutionary adaptation that allows animals to handle all the information that they receive from the sensory organs much more efficiently than if they were just relying on automatic reflexes or "instincts." Christoff Koch, a neuroscientist in Seattle, took us one step further when he wants to claim that single cells are conscious, as they can process information to their advantage. Thus, it is obvious that before having a meaningful discussion of consciousness, we would need to agree on some operational definition of the word.

It is difficult to deny that thinking is a result of processes in the brain. But what path would take us from two individual nerve cells exchanging signals to our brains reading and understanding a Tolstoy novel or planning our Christmas gift purchases? Does the brain resemble a computer? It happens to be the best metaphor we have at the moment—we only have limited knowledge of how the brain handles information, in what form our memories are stored, or if there is any kind of "program" in the brain. 17th century philosophers wondered if thinking arose through mechanical processes. The influential German 19th century physicist Hermann von Helmholtz suggested in his time that the brain could be compared to a telegraph. Each epoch finds its models.

One can argue that the cognitive ability of our brain is so complex that a classical supercomputer cannot even simulate it in principle. In that case, even a classical artificial intelligence—an AI—cannot under any circumstances develop artificial self-consciousness. But there are scientists who claim that the brain uses something more powerful than classical physics—which is why some people introduce quantum physics and what we might call "quantum consciousness" into the arena. So, what do we know today? Does consciousness arise through invisible events at the quantum level in the brain? Is the brain a quantum computer?

Quantum computers have been a concept at least since the American physicist Richard P. Feynman in 1982 presented the idea as a theoretical possibility. After all, there are problems that are too advanced for an ordinary classic computer that works with simple ones and zeros to be able to perform, calculations where the possible outcomes are astronomically large. In a quantum computer, the computational ability grows exponentially, which means that you can quickly get into precisely astronomically large numbers, thanks to the peculiarities of quantum mechanics.

A quantum bit—qubit—is an object that can be described in terms of quantum states. For example, it may be the polarization states of a photon, or the spin states of an electron, or the energy levels of an atom. All these states have definite values of the type "either-or" when observing them by a measurement. The spin is either up or down, the polarization is either right or left, the energy levels are at certain steps and so on. This means that these values are excellent digital information carriers, like the ones and zeros in a regular computer.

However, unlike the bits in a classical computer, the qubits can have more than one value at a time—as we have described the quantum physical principles. One bit, the basic unit of information, in a regular computer is either one or zero. In the quantum computer, it can be both one and zero at the same time—the qubit can be in *a superposition of zero and one*. So, while one hundred bits in a classic computer is a row with one hundred ones and zeros, one hundred qubits in a quantum computer corresponds to a cloud of possible outcomes consisting of 2^{100} possible answers. It is a number with 30 zeros—greater than the number of seconds that have elapsed since the Big Bang, when the universe was created over 13 billion years ago. So, you quickly arrive at incomprehensibly large numbers.

Quantum computers can make certain types of calculations much faster than classical computers and can store much more information in memory. Quantum computers could thus solve very difficult complex problems. This is one reason why much work has been put into explaining (or rather speculating on) how the brain could use quantum physics for information processing.

A well-known speculation for quantum consciousness was presented in the 1990s by the British cosmologist Sir Roger Penrose and the American anesthesiologist Stuart Hameroff. Penrose received the Nobel Prize in 2020, for proving the existence of black holes in Einstein's general theory of relativity and he has also collaborated with Steven Hawking. Together, Penrose and Hameroff developed their so-called "Orchestrated Objective Reduction" or OrchOR theory. The hypothesis is based on the network of long narrow tubes of protein molecules, microtubules, that form a kind of internal skeleton in all cells, and therefore also in the nerve cells in the brain.

OrchOR states, simply expressed, that electrons in the molecules make these narrow tubes vibrate in a way that affects the electrical and magnetic activity of the brain. The vibrating currents generated can have two different values and are assumed to be in superposition, that is in the two states simultaneously.

The bundles of these microtubules then form a system of qubits that can store and process information—as well as the quantum computer's ability for gigantic computations—and thus create our consciousness, according to Penrose and Hameroff. The brain is constantly vibrating with quantities of quantum information, one could say, according to this model.

The idea has become popular with some philosophers and spiritually influenced thinkers—but neuroscientists and physicists are very skeptical. The Swedish physicist Max Tegmark, who works at MIT in Boston, already wrote in the year 2000 an article that pointed out that electronic quantum processes in the microtubules cannot work as the OrchOR model requires.

Admittedly, new experiments have been done that show that there are interesting long-lived vibrational effects in microtubules. But this does not change anything fundamentally; that our cognitive ability depends on quantum mechanical effects in the nerve cells remains pure speculation.

But there are other hypotheses.

What happens, for example, if we instead assume that our thinking is influenced by the phenomenon that we talked about in Chapter 14, and which is the basis for the MR-camera's ability to see into the body's innermost parts? Involving the ability to create resonance among nuclear spins so that many particles "oscillate together" and emit weak electromagnetic radiation.

A prerequisite for MRI is that our body, not least our brain, is full of molecules with nuclear spin. Nuclear spins can have very long coherence times that persist in wet and warm environments such as brain nerve cells. NMR is used to create images of what happens in the brain when we perform various tasks. Can molecular quantum bits in the form of nuclear spins start to interact with each other, and affect the signaling in the brain?

This is in fact an idea put forward by Matthew Fisher, a theoretical physicist at the Kavli Institute in Santa Barbara, California. Fisher is trying to find biocompatible components that could perform quantum calculations.

In some way, according to theory, it would allow the brain to function at high speed and even make a distinction between consciousness and non-consciousness. But one might wonder why consciousness would arise in a brain that is a quantum computer? Nothing supports this.

Consciousness is notoriously difficult to define, but it is not primarily determined by the ability to handle enormous amounts of information. It's about something else. The difficulty lies in describing "spiritual values" as feelings, intentions, personal perception of colors and much more—Chalmers' hard problem. Many philosophers maintain the notion that emotions cannot be described by algorithms, and they cannot be described and simulated by computers—an AI is simply not considered capable of displaying emotions in the same way as a human being. It's not that it takes too long for a supercomputer to describe emotions—it's simply considered algorithmically impossible. Even for a quantum computer.

For many natural scientists and computer scientists, however, such arguments about emotions and algorithms feel like more faith than knowledge. We believe we understand, in principle, how the brain works, and one can simulate the behavior of several of the brain's networks with the help of powerful computers. Of course, subjective emotions and self-awareness do not arise by themselves— they are dependent on all the impressions and training the brain has experienced through its human body. But that information is real, physical, stored in memories and processed by neural networks. In principle, it should therefore be possible to describe it in terms of algorithms that can be run on supercomputers. But such things are certainly quite far into the future, and maybe not even practical—the brain simply works much smarter and more efficiently than that.

However, if one, after all, considers consciousness to be a function that arises in our brains and is not a soul-like phenomenon, no spirit separated from body and matter, then one should perhaps assume a reasonable classical model for how thinking works, and then possibly explore how that simpler classical model would be refined using quantum physics. And we still largely miss even the basic classical model to start from.

With modern techniques we can study the activity of the brain in real time—the question then is what it tells us about consciousness, and at what level self-awareness arises. Where is the center of control that links our impressions of mind to images of the world and makes us perceive happiness and grief?

One can map which parts of the brain are active when a subject does different things. For example, lying and thinking and daydreaming, or reacting to commands with movements. In this way, a network of connections between different regions of the brain has been identified, which is mainly active when we lie perfectly still and daydreaming, only aware of ourselves—the resting network. This network shuts down when the subject is forced to respond to external or internal activity that requires attention. Then other networks are activated instead that control responsiveness and motor skills—thus more reflexive behavior. It seems reasonable to associate the resting network with cognition and self-awareness—with the "self." It is also the resting network that responds most strongly to psychedelic drugs such as LSD and ecstasy. The effect has been described as "the self dissolves."

What has been observed in these studies of the brain's various networks suggests that the brain is not a computer in the ordinary sense, but that it does well with classical physics. One does not need to introduce speculative quantum physical assumptions to understand it. Today, we can study the activity in different parts of the brain in real time with the aid of a number of advanced tools based on nano- and quantum technologies. With these modern tools, we can now observe and distinguish the variation of brain patterns with time and relate them to documented activities, like moving an arm or looking at the faces of different people. But it does not stop here—if one can measure signals, then one can also use them to achieve some action. And perhaps even create feedback to influence or control the source of the signals. There are many examples of humans and animals being able to control external devices like computer cursors or robot arms "just by thinking about it."

This has led to the fascinating emerging field of Brain-Computer Interfaces (BCI), often also referred to as Brain-Machine Interfaces (BMI). BCI has been a hot topic for decades, but the recent advances in measurement technology, data processing, machine learning and visualization promise to create new possibilities that sound like science fiction. By implanting sensor arrays under the skull, on top of selected regions of the cortex, one can get information about intended and actual motor actions and control prosthetic devices. By analyzing the time patterns in EEG signals recorded via a helmet with a dense array of electrodes, one can decide whether a person is in a vegetative state or "just" paralyzed. If paralyzed, the person may then still be able to communicate and experience at least a little bit of a normal intellectual life.

Of course, this is by no means the same as being able to read someone's thoughts, but it does indicate that it may become possible in the future at least to decode simple thoughts in rough terms.

In the introduction to our book, we noted that our knowledge of the universe and matter has gone far beyond the bounds of what human senses can perceive. But the biggest mystery of them all, seen from our narrow human perspective, is the body of a little over a kilogram of nerve cells that we carry under the crown of our heads— we still understand very little of it so far. Future generations will probably understand largely the principles of the biological brain as the advanced operator it is. But creating algorithms and robots with AI and "real human emotions" probably remains science fiction for quite some time.

There are those who claim that we humans have no free will. Instead, we would be guided by algorithms: a kind of equivalent to computer programs.

It may be true that the basic functions of the brain's network can in principle be simulated by a classical computer. What is required for such a simulation to develop a free will is that the program is not locked into the individual's history, that it can move beyond the frameworks that have been set up and the experiences it has made during its life journey. It must be able to jump between different possibilities, new unknown alternatives and especially those that can be perceived as free fantasies—making decisions that are completely unique.

Those who argue that it is quantum physics that govern the brain often link these jumps to random quantum jumps—the unpredictable outcomes that result from Heisenberg's uncertainty principle. The idea connects quantum physical descriptions of matter with our inner worlds of thought in a way that many find appealing. However, we who argue for a classical brain mean that there are enough random processes in our complex biological systems to give us free will. How we use it, though, is a different matter.

Quantum physics appeals to the imagination of those who believe in supernatural phenomena. For many, it becomes natural to add the word "quantum" as a prefix to a variety of phenomena to reinforce the message or to explain invalid claims. And not least for marketing various products. We live in a time where deliberate disinformation is part of everyday life. It is dangerous for a democratic society if its citizens base their actions on illusions. So nowadays, and in the future, it is important to be enlightened and to be able to distinguish between fact and fantasy.

The enormous technological development of the last 50 years has created an abyss between what we ordinary people understand, and what we actually have access to and utilize and take for granted, in our everyday lives. The ever-increasing computing capacity of classical computers already makes it possible to create self-learning apps—programs that themselves create different forms of AI. Quantum computers and quantum physics will only help to amplify that development.

Soon we are dealing with smart artificial systems that give us good, or bad, advice in all possible contexts. We will then need a fair amount of knowledge—and maybe also common sense—to be able to evaluate the information that flows past us. We hope that this book can help broaden the perspective slightly on the road toward an exciting and challenging future.

Chapter 17

A Glance at the Future of Quantum and Life

Natural science does not simply describe and explain nature; it is part of the interplay between nature and ourselves.

—Werner Heisenberg, physicist and quantum physics explorer

We don't know exactly how life once began, but we can speculate and draw up more or less probable scenarios. The same goes for the future. Where will quantum biology, quantum technology, and, in the long run, even Evolution take us?

It is now about 80 years left to the turn of the century. A safe bet is that houses, cities, the sun, and our Milky Way galaxy will roughly look like they do today. But what about everything else? It is difficult to predict where our curiosity will bring us up to the turn of the century—200 years after the birth of quantum physics. Hopefully, it will only kill Schrödinger's cat, not us. In cosmology there is the never-ending quest for the "theory of everything," unifying all forces, describing the Big Bang, and explaining the Evolution at all levels.

But most fundamentally for our discussion—how exactly do quantum effects influence biological systems? 100 years of quantum experience tells us that this will take some time to answer. Even if one has the correct ideas, it may take a long time to prove them in principle. And it may take much longer to establish, or disprove, them

Quantum Physics and Life
Ingemar Ernberg, Göran Johansson, Tomas Lindblad, Joar Svanvik, and Göran Wendin
Copyright © 2023 Jenny Stanford Publishing Pte. Ltd.
ISBN 978-981-4968-28-7 (Hardcover), 978-1-003-31267-3 (eBook)
www.jennystanford.com

with certainty because a correct evaluation of the existing theory might need advanced computational schemes and top-performance supercomputers.

This is particularly visible in discussions of the role of quantum coherence in biological systems—the main theme of this book. What is it that cannot be described by classical physics and biochemical reactions? We have shown several examples, but how many more are there? How will the fast technical evolution, with quantum physics as a vital ingredient, change the view of biology and its relation to physics? And how will it influence medicine and health care?

To get some perspective on what the possibilities and challenges might look like, let's go back 80 years to the 1940s. Those days saw the start of a technological revolution with the birth of radar, transistors, quantum electronics, computers, molecular biology, and neurobiology. The 1950s saw the advent of protein crystallography, the discovery of the alpha-helix structure of proteins and the DNA double helix, and the first use of NMR in medicine. In the 1960s came lasers, integrated circuits, "supercomputers," computer modelling of materials, computer communication networks, the first satellite Sputnik, and the Apollo moon project.

But maybe the first tipping point was already in 1959 with the birth of semiconductor microtechnology, initiating the IT revolution. Then the first integrated semiconductor circuit was invented, and Richard Feynman delivered his famous speech at Caltech claiming that "There is plenty of room at the bottom," often described as the theoretical birth of nanotechnology.

At that time, very few people could imagine how these ideas would affect us in just a few decades—in particular the fact that the computational capacity of supercomputers and algorithms would increase exponentially.

The consequence is that we are now beginning to be able to create and simulate computer-driven complex systems with emergent behavior. We now have tools to influence biological and societal evolution. We now have the tools to dramatically expand our senses and functionality as human beings.

The figure at the beginning of this book and Chapter 2 describe how our biological senses and abilities are limited by nature, and how we extend ourselves using technology. We are now beginning to explore life, Earth, and the universe with present and future classical and quantum tools—a (hopefully) never-ending story.

The exponential development of the life sciences with molecular biology, gene editing, and high-resolution microscopy will enhance our insights and understanding of biological function. Quantum biology in turn is based on quantum chemistry and quantum mechanics. The insight into quantum chemistry will revolutionize our understanding of plant and animal biochemistry. Furthermore, understanding light-harvesting systems based on photosynthesis will help us to build efficient systems for harvesting energy from the sun. Understanding the role of quantum coherence in biology allowing tunneling, spin correlation and entanglement will give us new insights into enzymatic and biological functions.

There is talk about "the singularity," the notion that a computer may soon, in some sense have higher data processing capacity than our brains. However, visions about computers mimicking the brain miss the point by a wide margin. What is really important in a "short" 25-year perspective is AI based on *artificial neural networks*, ANN, deep neural networks, and machine learning.

ANNs are often referred to as "brain-inspired," but mimic at best highly specialized circuits in real brains, like the neural nets processing information from the eye's retina. In practice, ANNs are computer programs—algorithms—that filter information. Image processing with deep neural networks is a typical application: all the pixels in the original image are grouped in larger and larger chunks through a sequence of layers of neurons, and there is feedback from the back toward the front layers ("backpropagation") to tune the synaptic weights of the links. In the end, the output neurons present "a picture of the image."

That may sound trivial, but it certainly is not. A camera takes a photo and presents us with an image. But that image means nothing to us unless we understand how to interpret it. Our brains have been trained to do so over the years through everyday experience. But when a doctor looks at an X-ray picture of your lungs or inspects a picture of some area of your skin, there is a lot of image processing that must take place in the doctor's brain in order to conclude that this is cancerous tissue that has to be removed.

Here is one instance where the huge increase in computer processing power becomes important. AI with ANNs was introduced already in the 1960s but became regarded in the 1990s as a failed academic invention until powerful computers ten years later made them useful beyond imagination. (A good example of the unpredictability of technical progress. Which of today's failures will revolutionize the future? And what hyped modern technology will be forgotten in a generation?)

The bottom line is that ANNs can be trained to store and represent large amounts of data, and the computer AI can then search for patterns in those databases. This is how AI can identify, for example, skin cancer much more reliably than most trained medical doctors. And predict how a protein will fold more efficiently than the best mathematical algorithms.

So, what about quantum computers? They are certainly here to stay. But the question is to what extent quantum technologies will have revolutionized the world already by 2045. Today's quantum computers and simulators are currently pretty much in the same situation as were classical vacuum tube computers in the 1940s—constantly exposed to errors and failures. This time, however, the quantum evolution will be faster, and the current roadmaps aim for useful quantum computers around 2030. Nevertheless, this does not mean that quantum computers will replace classical supercomputers—they will become integrated as accelerators to speed up computations where there is proven quantum advantage or quantum superiority.

In the future, quantum technology will most certainly influence various areas of "digital" health care, making use of nanotechnology, biochemistry, high-resolution microscopy, and imaging for diagnosis for preventive treatment and therapy, often via remote control.

Living cells consist of conglomerates of natural functional nanoparticles that interact with each other and the environment by exchanging chemicals transported by nanoparticles like lysosomes. Viruses are natural nanoparticles that can attach to a cell and prompt it to open its doors.

Artificial functionalized nanoparticles work along the same principles, and they have become very important tools for diagnosis and therapy. Quantum dots are made of small semiconductor or metallic grains decorated with a fur of molecules with different functions that can attach to cells and make them accessible for radiation treatment. Other nanoparticles are functionalized hollow containers similar to viruses or lysosomes and can deliver drugs to desired targets. For example, the mRNA vaccines that were developed to protect from Covid-19 are packaged in a protein container much like a virus.

Quantum chemistry, AI and machine learning will help us design new molecules. Already today they can predict how a given chain of amino acids will fold into a protein, providing better results than the best computational schemes. With the capacity to compute the molecular vibrations of chemical bonds, the binding of molecules to cellular receptors can be better predicted and make the effects of new drugs more specific.

Analyzing cancers in a patient will inform oncologists how to better individualize treatment. This involves tuning immune systems, vaccines, antibiotics, and multi-resistant bacteria, as well as phages.

One development that may be influential already by 2050 is the consequences of imaging the brain. New techniques make it possible to see and interpret the activities of the different brain networks and implants for brain-controlled devices; brain-controlled prosthetics; brain-controlled brains; even decoding brains—to some extent revealing thoughts, internal images and maybe even feelings.

Prophylactic measures and early diagnosis of disease using MRI technologies will refine medicine: for example, AI-controlled personalized diagnoses, cure of cancer with therapies based on detailed characterization of immune systems, design of vaccines, antibiotics, bacteria, and viruses with the help of gene engineering. Quantum chemistry supported by machine learning will help us develop more specific ligands for receptors. Tuning molecular vibrations in chemical bonds will make ligand-receptor-binding more specific and the pharmacological effects more efficient. Quantum dots and vesicles functionalized with highly specific ligands can then improve the precision of tools for therapy, for instance carrying toxins to selected targets. Advanced measurement technologies and computational biology characterizing specific cancers in individuals will inform oncologists how to better personalize the treatment.

But what about *quantum coherence in biological systems*, the central question for quantum biology? Can we imagine that we will be able to control and make use of quantum coherence and entanglement to improve medical diagnosis and treatment, and even to correct some of nature's "mistakes"?

When it comes to diagnostic tools, the answer is—yes. Quantum sensors, like functionalized quantum dots or color centers in diamond nanoparticles, will make use of coherence and entanglement to increase the sensitivity of detection and the resolution of medical images. Quantum sensors are effectively quantum bits, so in future, one may be probing the interior of biosystems with nanoscale quantum computers!

When it comes to curing diseases, the answer is—no. Or better said—the idea is not useful. Quantum coherence is already a fundamental property of small biomolecules. Boosting quantum coherence in some part of a large molecule will simply modify the biochemical properties of that part. This may indeed modify the folding of a protein or increase the energy transfer between two proteins. And, in principle, this might correct misfolding of proteins, or modify enzymes and change the metabolism of biological cells. However, *the outcome is beyond human control*. The only way to permanently change the properties of proteins in living tissue is to genetically modify the expression of specific genes. But that is surely beyond the detailed control of medical science and cannot serve as therapy—the biochemical effects on complex biological networks are simply *not predictable*. Evolution knows this because it has been experimenting with these things for billions of years.

In the end, maybe the *future of Life* may be about quantum physics. Mutations, whether due to chance, radiation, or reactive aggressive molecules, change DNA at the quantum level of the basic code letters—C : G or A : T. So far, nature has developed repair mechanisms that have taken us up to the present time. However, now we humans can identify any faulty DNA sequences and correct them. We now know the secrets of bacterial defenses and may be able to use them to change our existence.

"What is life?" Erwin Schrödinger asked in 1944. This is indeed a big question, and in this book, we have tried to provide a simple guide to some answers. It is also a very serious question, so serious that one must not take it too seriously. If you do, you will most often come up with answers that convince no one but yourself. But it surely leads to interesting discussions and hot debates.

References and Recommended Readings

Chapter 1. Life and Quantum Physics

Arndt M., Juffmann T., Vedral V. (2009). Quantum physics meets biology. *HFSP J.*, **3**, 386–400.

Lloyd S. (2011). Quantum coherence in biological systems. *J. Phys.*, **302**, 012037.

Ball P. (2011). Physics of life: the dawn of quantum biology. *Nature*, **474**, 272–274.

Li C. M., Lambert N., Chen Y. N., Chen G. Y., Nori F. (2012). Witnessing quantum coherence: from solid-state to biological systems. *Sci. Rep.*, **2**, 885.

Lambert N., Chen Y.-N., Cheng Y.-C., Li C.-M., Chen G. Y., Nori F. (2013). Quantum biology, *Nat. Phys.*, **9**, 10–18.

Al-Khalili J., McFadden J. (2014). *Life on the Edge: The Coming of Age of Quantum Biology*. London, UK: Bantam Press.

Brookes J. C. (2017). Quantum effects in biology: golden rule in enzymes, olfaction, photosynthesis and magnetodetection. *Proc. Math. Phys. Eng. Sci.*, **473**, 2016082.

Marais A., *et al.* (2018). The future of quantum biology. *J. R. Soc. Interface*, **15**, 20180640.

McFadden J., Al-Khalili J. (2018). The origins of quantum biology. *Proc. Math. Phys. Eng. Sci.*, **474**, 20180674.

Cao L., et *al.* (2020). Quantum biology revisited. *Sci. Adv.*, **6**, eaaz4888.

Kim Y., et al. (2021). Quantum biology: an update and perspective. *Quantum Rep.*, **3**, 80–126.

Chapter 2. Our World is Just One Part of the Whole

Nagel T. (1974). What is it like to be a bat? *Philos. Rev.*, **83**, 435–450.

Monarch Butterflies Migrate 3,000 Miles—Here's How. (2017). National Geographic. https://www.nationalgeographic.com/animals/article/monarch-butterfly-migration

Thogmartin W. E., *et al.* (2017). Monarch butterfly population decline in North America: identifying the threatening processes. *R. Soc. Open Sci.*, **4**, 170760.

Hubble Space Telescope/NASA

https://www.nasa.gov/mission_pages/hubble/main/index.html

Hubble Space Telescope: Pictures, Facts & History

https://www.space.com/15892-hubble-space-telescope.html

LHC (Large Hadron Collider). https://home.cern/science/accelerators/large-hadron-collider

Schrödinger E. (1944). What Is Life? The Physical Aspect of the Living Cell. Based on Lectures delivered under the auspices of the Institute at Trinity College, Dublin, in February 1943. Cambridge: University Press. 1944.

Was ist Leben? Die lebende Zelle mit den Augen des Physikers betrachtet [Translation into German by L. Mazurczak]. Bern: Francke. 1946. München: Lehnen. 1946.

Qu'est-ce que la vie? L'aspect physique de la cellule vivante. [Translation into French by L. Keffler]. Paris: Club français du livre, 1949.

Chapter 3. The Gecko and Life Upside Down

Persson B. N. J. (2003). On the mechanism of adhesion in biological systems. *J. Chem. Phys.*, **118**, 7614–7621.

Song Y., Wang Z., Zhou J., Li Y., Dai Z. (2018). Synchronous measurement of tribocharge and force at the footpads of freely moving animals. *Friction*, **6,** 75–83.

Shuttleworth I. G. (2020). On the role of the van der Waals interaction in gecko adhesion: a DFT perspective. *Results Mater.*, **6**, 100080.

Mahdavi A., et al. (2008). A biodegradable and biocompatible gecko-inspiredtissue adhesive. *PNAS*, **105**, 2307–2312.

Chapter 4. The Quantized World

Schrödinger E. (1926). *Quantisierung als Eigenwertproblem* (*Erste Mitteilung*), *Annalen der Physik*, **79**, 361–376; 489–527; **80**, 437–490.

Schrödinger E. (1926). An undulatory theory of the mechanics of atoms and molecules. *Phys. Rev.*, **28**, 1049–1070.

Schrödinger E. (1926). *Über das Verhältnis der Heisenberg-Born-Jordanschen Quantenmechanik zu der meinen. Annalen der Physik*, **79**, 734–756.

Schrödinger E. (1932). *Über Indeterminismus in der Physik—Ist die Naturwissenschaft milieubedingt? Zwei Vorträge zur Kritik der naturwissenschaftlichen Erkenntnis.* Leipzig: Barth.

Schrödinger E. (1936). Indeterminism and free will. *Nature*, **138**, 13–14.

Schrödinger E. (1937). World structure. *Nature*, **140**, 742–744.

Heisenberg W. (1925). *Über quantentheoretische Umdeutung kinematischer und mechanischer Beziehungen. Z. Phys.*, **33**, 879–893.

Born M., Jordan P. (1925). *Zur Quantenmechanik. Z. Phys.*, **34**, 858–888.

Born M., Heisenberg W., Jordan P. (1926). *Zur Quantenmechanik II. Z. Phys.*, **35**, 557–615.

Dirac P. A. M. (1926). The fundamental equations of quantum mechanics. *Proc. Roy. Soc. A*, **109**, 642–653.

Tonomura A., Endo J., Matsuda T., Kawasaki T. (1989). Demonstration of single-electron buildup of an interference pattern. *Am. J. Phys.*, **57**, 117–120.

Bach R., Pope D., Liou S.-H., Batelaan H. (2013). Controlled double-slit electron diffraction. *New J. Phys.*, **15**, 033018.

Chapter 5. Evolution: About the Origin of Life

Darwin C. (1859). *On the Origin of Species by Means of Natural Selection, or the Preservation of Favoured Races in the Struggle for Life* (1st Ed.). London: John Murray, p. 502, retrieved 1 March 2011.

Dodd M. S., *et al.* (2017). Evidence for early life in Earth's oldest hydrothermal vent precipitates. *Nature*, **543**, 60–64

Damer B., Deamer D. (2020). The hot spring hypothesis for an origin of life. *Astrobiology*, **20**, 429–452.

Trapp O., Teichert J., F. Kruse. (2019). Direct prebiotic pathway to DNA nucleosides. *Angewandte Chem.* (International Edition), **58**, 9944–9947.

Kriegman S., Blackiston D., Levin M., Bongard J. (2020). A scalable pipeline for designing reconfigurable organisms. *PNAS*, **117**, 1853–1859.

Kauffman, S. A. (2019). *A World Beyond Physics: The Emergence and Evolution of Life*. New York, USA: Oxford University Press.

Petrović D., Risso V. A., Kamerlin S. C. L., Sanchez-Ruiz J. M. (2018). Conformational dynamics and enzyme evolution. *J. R. Soc. Interface*, **15**, 20180330.

Nagel Z. D., Klinman J. P. (2006). Tunneling and dynamics in enzymatic hydride transfer. *Chem. Rev.*, **106**, 3095–3118.

Lu J., Luo L. (2014). Statistical analyses of protein folding rates from the view of quantum transition. *Sci. China Life Sci.*, **57**, 1197–1212.

Kamerlin S. C., Warshel A. (2011). Multiscale modeling of biological functions. *Phys. Chem. Chem. Phys.*, **13**, 10401–10411.

Söderhjelm P., Kongsted J., Genheden S., Ryde U. (2010) Estimates of ligand-binding affinities supported by quantum mechanical methods. *Interdiscip. Sci.*, **2**, 21–37.

Dawkins R. (1989). *The Selfish Gene* (2nd Ed.). New York, USA: Oxford University Press.

Wu L. F., Sutherland J. D. (2019). Provisioning the origin and early evolution of life. *Emerg. Top. Life Sci.*, **3**, pp. 459–468.

Santiago-Alarcon D., Tapia-McClung H., Lerma-Hernández S., Venegas-Andraca S. E. (2020). Quantum aspects of evolution: a contribution towards evolutionary explorations of genotype networks via quantum walks. *J. R. Soc. Interface*, **17**, 20200567.

Preiner M., *et al.* (2020). The future of origin of life research: bridging decades-old divisions. *Life*, **10**, 20.

Chapter 6. From the Big Bang to Black Holes

Freeth T., *et al.* (2021). A model of the cosmos in the ancient Greek antikythera mechanism. *Sci. Rep.*, **11**, 5821.

Weinberg S. (1993). *The First Three Minutes: A Modern View of the Origin of the Universe.* New York, USA: Basic Books. ISBN-13: 9780465024377.

Al-Khalili J. (2020). *The World According to Physics.* Oxfordshire, UK: Princeton University Press.

Sokol J. (2020). Physicists argue that black holes from the big bang could be the dark matter. *Quantamagazine*.

https://www.quantamagazine.org/black-holes-from-the-big-bang-could-be-the-dark-matter-20200923/

Yu H. and other members of the LIGO Scientific Collaboration. (2020). Quantum correlations between light and the kilogram-mass mirrors of LIGO. *Nature,* **583**, 43–47.

Chapter 7. As Time Goes By: The Arrow of Time

Loudon A. S. I. (2012). Circadian biology: a 2.5 billion year old clock. *Curr. Biol.,* **22**, R570.

Buzsáki G., Llinás R. (2017). Space and time in the brain. *Science*, **358**, 482–485.

Buzsáki G., Tingley D. (2018). Space and time: the hippocampus as a sequence generator. *Trends Cogn. Sci.,* **22**(10), 853–869.

Tsao A., Sugar J., Lu L., Wang C., Knierim J. J., Moser M.-B., Moser E. I. (2018). Integrating time from experience in the lateral entorhinal cortex. *Nature*, **561**, 57–62.

Shapiro M. (2019). Time is just a memory. *Nat. Neurosci.,* **22**, 151–153.

Chapter 8. The Art of Finding Your Way Home

Ritz T., Adem S., Schulten K. (2000). A model for photoreceptor based magnetoreception in birds. *Biophys. J.,* **78**, 707–718.

Ritz T., Yoshii T., Helfrich-Foerster C., Ahmad M. (2010). Cryptochrome: a photoreceptor with the properties of a magnetoreceptor? *Commun. Integr. Biol.,* **3**, 24–27.

Heyers D., Manns M., Luksch H., Gunturkun O., Mouritsen H. (2007). A visual pathway links brain structures active during magnetic compass orientation in migratory birds. *Plos One*, e937.

Rodgers C. T., Hore P. J. (2009). Chemical magnetoreception in birds: the radical pair mechanism. *PNAS*, **106**, 353–360.

Lau J. C., Rodgers C. T., Hore P. J. (2012). Compass magnetoreception in birds arising from photo-induced radical pairs in rotationally disordered cryptochromes. *J. R. Soc. Interface*, **9**, 3329–3337.

Cai J. and Plenio M. B. (2013). Chemical compass model for avian magnetoreception as a quantum coherent device. *Phys Rev. Lett.,* **111**, 230503.

Binhi V. N. (2019). Nonspecific magnetic biological effects: a model assuming the spin-orbit coupling. *J. Chem. Phys.*, **151**, 204101.

Chapter 9. The Vision in New Light

Sia P. I., Luiten A. N., Stace T. M., Wood J. P., Casson R. J. (2014). Quantum biology of the retina. *Clin. Exp. Ophthalmol.*, **42**, 582–589.

Tinsley J. N., Molodtsov M. I., Prevedel R., Wartmann D., Espigulé-Pons J., Lauwers M., Vaziri A. (2016). Direct detection of a single photon by humans. *Nature Commun.*, **7**, 12172.

Caprara Vivoli V., Sekatski P., Sangouard N. (2016). What does it take to detect entanglement with the human eye? *Optica*, **3**, 473–476.

Dodel A., Mayinda A., Oudot E., Martin A., Sekatski P., Bancal J.-D., and Sangouard N. (2017). Proposal for witnessing non-classical light with the human eye. *Quantum*, **1**, 7.

Chapter 10. The Photosynthesis and the Golf Putt

Engel G. S., Calhoun T. R., Read E. L., Ahn T.-K., Mancal T., Cheng Y.-C., Blankenship R. E., Fleming G. R. (2007). Evidence for wavelike energy transfer through quantum coherence in photosynthetic systems. *Nature*, 446, 782–786.

Chan H. C. H., Gamel O. E., Fleming G. R., Whaley K. B. (2018). Single-photon absorption by single photosynthetic light-harvesting complexes. *J. Phys. B*, **51**, 054002.

Duan H.-G., *et al.* (2017). Nature does not rely on long-lived electronic quantum coherence for photosynthetic energy transfer. *PNAS*, **114**, 32, 8493–8498.

Cao L., et *al.* (2020). Quantum biology revisited. *Sci. Adv.*, **6**, eaaz4888.

Chapter 11. The Respiratory Chain Sustains Our Lives

Moser C., Tammer A., Chobot S., Dutton L. (2006). Electron tunnelling of mitochondria. *Biochim. Biophys. Acta.*, **1757**, 1096–1109.

Lucas M. F., Rousseau D. L., Guallar V. (2011). Electron transfer pathways in cytochrome c oxidase. *Biochim. Biophys. Acta.*, **1807**, 1305–1313.

Blomberg M. and Siegbahn P. (2012). The mechanism for proton pumping in cytochrome c oxidase from an electrostatic and quantum chemical perspective. *Biochim. Biophys. Acta.*, **1817**, 495–505.

Leonid A. Sazanov L. A. (2015). A giant molecular proton pump: structure and mechanism of respiratory complex I. *Nat. Rev. Mol. Cell Biol.,* **16**, 375–388.

Cogliati S., Enriquez J. A., Scorrano, L. (2016). Mitochondrial cristae: where beauty meets functionality. *Trends Biochem. Sci.*, **41**, 261–273.

Weinrich T. W., Kam J. H., Ferrara B. T., Thompson E. P., Mitrofanis J., Jeffery G. (2019). A day in the life of mitochondria reveals shifting workloads. *Sci. Rep.*, **9**, 13898.

Caruana N. J., Stroud D. A. (2020). The road to the structure of the mitochondrial respiratory chain supercomplex. *Biochem. Soc. Trans.*, **48**, 621–629.

Marušič N., *et. al.* (2020). Constructing artificial respiratory chain in polymer compartments: Insights into the interplay between bo3 oxidase and the membrane. *PNAS* **117**, 15006–15017.

Chapter 12. A Sense of Smell

Dyson, G. M. (1938). The scientific basis of odour. *Chem. Ind.*, 647–651.

Lambe J. and Jaklevic R. C. (1968). Molecular vibration spectra by inelastic electron tunneling. *Phys. Rev.*, **165**, 821–832.

Turin, L. (1996). A spectroscopic mechanism for primary olfactory reception. *Chem. Senses*, **21**, 773–791.

Turin L., Yoshii F. (2003). Structure–odor relations: a modern perspective. In: Doty, R. L. (ed.), *Handbook of Olfaction and Gustation.* New York: Marcel Dekker, pp. 275–294.

Editorial. (2004). Testing a radical theory. *Nat. Neurosci.*, **7**, 315.

Keller A., Vosshall L. B. (2004). A psychophysical test of the vibration theory of olfaction. *Nat. Neurosci.*, **7**, 337–338.

Ball P. (2006). Rogue theory of smell gets a boost. *Nature.* https://doi.org/10.1038/news061204-10.

Turin L. (1996). A spectroscopic mechanism for primary olfactory reception. *Chem. Senses* **21**, pp. 773–791.

Brookes J. C., Hartoutsiou F., Horsfield A. P., Stoneham A. M. (2007). Could humans recognize odor by phonon assisted tunneling? *Phys. Rev. Lett.*, **98**, 038101.

Gane S., Georganakis D., Maniati K., Vamvakias M., Ragoussis N., Skoulakis E. M., Turin L. (2013). Molecular vibration-sensing component in human olfaction. *PLoS One*, **8**, e55780.

Bushdid C., Magnasco M. O., Vosshall L. B., Keller A. (2014). Humans can discriminate more than 1 trillion olfactory stimuli. *Science*, **343**, 1370–1372.

Turin L., Gane S., Georganakis D., Maniati K., Skoulakis E. M. (2015). Plausibility of the vibrational theory of olfaction. *PNAS*, **112**, E3154.

Hoehn R. D., Nichols D. E., Neven H., Kais S. (2018). Status of the vibrational theory of olfaction. *Front. Phys.*, **6**, 25.

Genva M., Kenne Kemene T., Deleu M., Lins L., Fauconnier M.-L. (2019). Is it possible to predict the odor of a molecule on the basis of its structure? *Int. J. Mol. Sci.*, **20**, 3018.

Chong E., Moroni M., Wilson C., Shoham S., Panzeri S., Rinberg D. (2020). Manipulating synthetic optogenetic odors reveals the coding logic of olfactory perception, *Science*, **368**, eaba2357.

Stetka B. (2020). The brain interprets smell like the notes of a song. *Scientific American* (18 June).

Sanchez-Lengeling B., Wei J. N., Lee B. K., Gerkin R. C., Aspuru-Guzik A., Wiltschko A. B. (2019). The chemistry of smell: learning generalizable perceptual representations of small molecules, *Second Workshop on Machine Learning and the Physical Sciences (NeurIPS 2019)*, Vancouver, Canada. Full paper: arXiv:1910.10685.pdf

del Mármol J., Yedlin M., Ruta V. (2021). The structural basis of odorant recognition in insect olfactory receptors, *bioRxiv preprint* doi: https://doi.org/10.1101/2021.01.24.427933.

Chapter 13. DNA Repair: Enzymes for Survival and Development

Sancar A. (1995). DNA repair in humans. *Annu. Rev. Genet.*, **29**, 69–105.

Hakem R. (2008). DNA-damage repair; the good, the bad, and the ugly. *EMBO J.*, **27**, 589–605.

Chatzidoukaki O., Goulielmaki E. Schumacher B., Garinis G. A. (2020). DNA damage response and metabolic reprogramming in health and disease, *Trends Genet.*, **36**, 777–791.

Sobanski T., Rose M., Suraweera A., O'Byrne K., Richard D. J., Bolderson E. (2021). Cell metabolism and DNA repair pathways: implications for cancer therapy. *Front. Cell Dev. Biol.*, **9**, 633305.

Lindahl T., Modrich P., Sancar A. (2015). DNA repair: providing chemical stability for life. Nobel Prize, Chemistry. https://www.nobelprize.org/uploads/2018/06/popular-chemistryprize2015.pdf

Chapter 14. Quantum Physics in Diagnosis and Treatment

Chan W. C., Maxwell D. J., Gao X., Bailey R. E., Han M., Nie S. (2002). Luminescent quantum dots for multiplexed biological detection and imaging. *Curr. Opin. Biotechnol.*, **13**, 40–46.

Smith A. M., Duan H., Mohs A. M., Nie S. (2008). Bioconjugated quantum dots for in vivo molecular and cellular imaging. *Adv. Drug Deliv. Rev.*, **60**, 1226–1240.

Pasparakis G. (2013). Light-induced generation of singlet oxygen by naked gold nanoparticles and its implications to cancer cell phototherapy. *Small*, **9**, 4130–4134.

Pasparakis G., Manouras T., Vamvakaki M., Argitis P. (2014). Harnessing photochemical internalization with dual degradable nanoparticles for combinatorial photo-chemotherapy. *Nat. Commun.*, **5**, 3623.

Kong K., *et al.* (2013). Diagnosis of tumors during tissue-conserving surgery with integrated autofluorescence and Raman scattering microscopy. *PNAS*, **110**, 15189–15194.

Singh P., *et al.* (2018). Gold nanoparticles in diagnostics and therapeutics for human cancer. *Int. J. Mol. Sci.*, **19**, 1979.

Zhang Y., Li M., Gao X., Chen Y., Liu T. (2019). Nanotechnology in cancer diagnosis: progress, challenges and opportunities. *J. Hematol. Oncol.*, **12**, 137.

Hansel C. S., Holme, M. N., Gopal, S., Stevens, M. M. (2020). Advances in high-resolution microscopy for the study of intracellular interactions with biomaterials. *Biomaterials*, **226**, 119406.

Tanner M. G., *et al.* (2017). Ballistic and snake photon imaging for locating optical endomicroscopy fibres, *Biomed. Opt. Express*, **8**, 4077–4095.

Barriga H. M. G., Holme M. N., Stevens M. M. (2019). Cubosomes: the next generation of smart lipid nanoparticles? *Ang. Chem.* (Int. Ed.), **58**, 2958–2978.

Werner M., *et al.* (2018). Nanomaterial interactions with biomembranes: bridging the gap between soft matter models and biological context. *Biointerphases*, **13**, 28501.

Grandin M., *et al.* (2018). Bioinspired, nanoscale approaches in contemporary bioanalytics (Review). *Biointerphases*, **13**, 040801.

Miller A. D. C., Ozbakir H. F., Mukherjee A. (2021). Calcium-responsive contrast agents for functional magnetic resonance imaging. *Chem. Phys. Rev.*, **2**, 021301.

Peer D., Karp J. M., Hong S., Farokhzad O. C., Margalit R., Langer R. (2020). Nanocarriers as an emerging platform for cancer therapy. In: Balog L. P. (ed.), *Nano-Enabled Medical Applications*. Singapore: Jenny Stanford Publishing, 61–91.

Casacio C. A., Madsen L. S., Terrasson A., Waleed M., Barnscheidt K., Hage B., Taylor M. A., Bowen W. P. (2021). Quantum-enhanced nonlinear microscopy. *Nature*, **594**, 201–206.

Chapter 15. No More Mysterious Than Necessary

Mermin N. D. (2019). Making better sense of quantum mechanics. *Rep. Prog. Phys.*, **82**, 012002.

Juan Yin, *et al.* (2020). Entanglement-based secure quantum cryptography over 1,120 kilometres. *Nature*, **582**, 501–505.

Gerlach W., Stern O. (1922). *Der experimentelle Nachweis der Richtungsquantelung im Magnetfeld. Zeitschrift für Physik.*, **9**, 349–352.

Tonomura A., Endo J., Matsuda T., Kawasaki T. (1989). Demonstration of single-electron buildup of an interference pattern. *Am. J. Phys.*, **57**, 117.

Schrödinger E. (1935). *Die gegenwärtige Situation in der Quantenmechanik* (The present situation in quantum mechanics). *Naturwissenschaften*, **23**, 807–812. Original paper introducing Schrödinger's cat.

Yin Z. and Li T. (2017). Bringing quantum mechanics to life: from Schrödinger's cat to Schrödinger's microbe. *Contemp. Phys.*, **58**, 119–139.

O'Callaghan J. (2018). "Schrödinger's bacterium" could be a quantum biology milestone. *Scientific American* (October 29, 2018).

Zurek W. H. (2002). Decoherence and the transition from quantum to classical: revisited. *Los Alamos Science* **27**, 2–25.

Zurek W. H. (2009). Quantum Darwinism. *Nat. Phys.*, **5**, 181–188.

Chapter 16. Consciousness: The Greatest Mystery

Nagel T. (1974). What is it like to be a bat? *Philos. Rev.*, **83**, 435–450.

Chalmers D. J. (1995). Facing up to the problem of consciousness. *J. Conscious. Stud.*, **2**, 200–219.

Horgan J. (2017). David Chalmers thinks the hard problem is really hard. *Scientific American* (April 2017).

Baars B. J., Edelman D. B. (2012), Consciousness, biology and quantum hypotheses. *Phys. Life Rev.*, **9**, 285–294.

Hameroff S. R., Craddock T. J., Tuszynski J. A. (2014). Quantum effects in the understanding of consciousness. *J. Integr. Neurosci.*, **13**, 229–252.

Tegmark M. (2015). Consciousness as a state of matter. *Chaos Solit. Fractals*, **76**, 238–270.

Wendin G. (2017). Can biological quantum networks solve NP-hard problems? *Adv. Quantum Technol.* **2**:1800081. arXiv:1902.03121v3.

Weingarten C. P., Murali D. P., Fisher M. P. A. (2016). A new spin on neural processing: quantum cognition. *Front. Hum. Neurosci.*, **10**, 541.

Swift M. W., Van de Walle C. G., Fisher M. P. A. (2017). Posner molecules: from atomic structure to nuclear spins. *Phys. Chem. Chem. Phys.*, **20**, 12373–12380.

Fisher, M. P. A., Radzihovsky L. (2018). Quantum indistinguishability in chemical reactions. PNAS, **115**, E4551–E4558.

Koch, C. (2018). What is consciousness? *Nature*, **557**, S9.

Chittka L, and Wilson C. (2018). Bee-brained. *Aeon* (ed. Sally Davies, 27 November 2018). https://aeon.co/essays/inside-the-mind-of-a-bee-is-a-hive-of-sensory-activity.

Chittka L., Wilson C. (2019). Expanding consciousness. *Amer. Sci*, **107**, 364–369.

Smith J., Zadeh-Haghighi H., Salahub D., Simon C. (2021). Radical pairs may play a role in xenon-induced general anesthesia. *Sci. Rep.*, **11**, 6287.

Zadeh-Haghighi H., Simon C. (2021). Entangled radicals may explain lithium effects on hyperactivity. *Sci. Rep.*, **11**, 12121.

Hoffmann C. (2003). Helmholtz' apparatuses telegraphy as working model of nerve physiology. *Philos. Sci.*, **7**, 129–149.

Gomez-Marin A. (2020). A history of the metaphorical brain. *Science*, **368**, 375.

Chapter 17. A Glance At the Future of Quantum and Life

Marcus G. (2017). Am I human? Researchers need new ways to distinguish artificial intelligence for the natural kind. *Scientific American (March 2017)*, 58–63.

Willett F. R., Avansino D. T., Hochberg L. R., Henderson J. M., Shenoy K. V. (2021). High-performance brain-to-text communication via handwriting. *Nature*, **593**, 249–254.

Cao Y., *et al.* (2019). Quantum chemistry in the age of quantum computing. *Chem. Rev.*, **119**, 10856–10915.

Charpentier E., Doudna J. A. (2020). Genetic scissors: a tool for rewriting the code of life. Nobel Prize in Chemistry 2020 for discovering one of gene technology's sharpest tools: the CRISPR/Cas9 genetic scissors.

https://www.kva.se/en/pressrum/pressmeddelanden/nobelpriset-i-kemi-2020

Miller A. D. C. , Ozbakir H. F., Mukherjee A. (2021). Calcium-responsive contrast agents for functional magnetic resonance imaging. *Chem. Phys. Rev.*, **2**, 021301

Johnson-Groh M. (2021). Magnetic resonance imaging of calcium ions could unlock the brain's mysteries. *AIP Scilight*. https://doi.org/10.1063/10.0004793

Patrick C. (2021). Nanoparticles labeled with antibodies hit their target. *AIP Scilight*. https://doi.org/10.1063/10.0004980

Kogos B., *et al.* (2021). Electron microscopy of antibody-conjugated, lutetium-177 lanthanide gold-coated nanoparticles: Proof of concept of targeted loci: a potential theranostic agent. *AIP Advances*, **11**, 045035.

Dictionary

Adipose tissue: Connective tissue involving fat cells.

Amino acid: Molecular building blocks of the cells' proteins. The general form of an amino acid molecule is $NH_2-CRH-COOH$. NH_2 is the amino group, COOH is the acid group, and R is a side group that characterizes the amino acid. Two amino acid molecules link together by the NH_2 group of one amino acid connecting to the COOH group of the other amino acid, forming a *peptide HN-CO bond* by splitting off a H_2O water molecule.

Atom: The basic building block of matter. An atom consists of a nucleus of protons and neutrons surrounded by a cloud of electrons.

Attosecond (as): 10^{-18} (0.000 000 000 000 000 001) seconds—a millionth of a picosecond. (See Table of Units).

Bit (computer): Information carrier that describes the logical levels 0 and 1. Bits code the information stored in a computer memory.

Bonding (molecules):
- *Covalent bond*: two or more atoms share electrons and these common electrons then "circle" around the atoms. For example, hydrogen (H_2), oxygen molecule (O_2), water (H_2O), hydrogen cyanide (HCN), formic acid (HCOOH).
- *Ionic bond*: one or more electrons have been completely transferred between atoms and created electrostatic attraction between positively and negatively charged ions. This is typical of salts, such as common salt, sodium chloride. This is often the case with the metal atoms found in the body's large biomolecules, such as the oxygen transporter hemoglobin with its iron atom in the center.

- *Hydrogen bond*: a hydrogen atom forms a molecular link by letting its electron tunnel between two molecules. A nice example is the attraction between two water molecules through a O–H–O hydrogen bond: H_2O–H–OH, giving water many of its unique properties.
- *Van der Waals bond:* The van der Waals force is an attractive force that acts between neutral (non-ionized) atoms and molecules at such great distances that electrons do not normally tunnel. The electromagnetic field leads to dipole polarization of the electron clouds which creates an attractive force at "long" distances.

Catalyst: A substance that speeds up chemical reactions without being consumed.

Cell membrane: A double layer of fatty acid molecules that forms the cell membrane.

Covalent bond: See Bonding (molecules).

Coherence: Characterizes an undisturbed, uninterrupted wave motion—a prerequisite for constructive and destructive interference. When coherence prevails, quantum effects can affect particle-waves in the microscopic world.

Cryptochrome: Light-sensitive pigment molecule that sits on the back of the retina of the eye, acts as a detector for photons and whose chemical properties can be affected by magnetic fields.

de Broglie wavelength, λ: $\lambda = h/mv$, where h is the Planck constant, m is the mass of the particle, and v is its speed.

Decoherence: Characterizes a disturbed, broken wave motion that blurs out interference patterns and thereby prevents large-scale effects of superposition.

DNA: Deoxyribonucleic acid. The molecule that carries the genetic information in the cell nucleus.

Electromagnetism (EM): The theory of wave motion in electric and magnetic fields caused by variable electric charges and electric currents that vary over time. Described mathematically by Maxwell's equations.

Electron: Elementary particle with negative charge –1.

Elementary particle: For the purpose of this book: protons, neutrons, electrons. At smaller scales, protons and neutrons are made up of quarks.

Energy: Ability to do work.

Entanglement: A symmetry quantum property connecting two, or many, quantum particles. The state of the particles cannot be described independently of each other.

Entropy: Measure of disorder.

Enzyme: A protein that acts as a catalyst.

Exciton: Electronic excitation where a negatively charged electron stays bound to the positively charged hole left behind.

Femtometer (fm): 10^{-15} (0.000 000 000 000 001) meters—a thousandth of a picometer. (See Table of Units).

Femtosecond (fs): 10^{-15} (0.000 000 000 000 001) seconds—a thousandth of a picosecond. (See Table of Units).

Free energy: Represents the part of energy that can do work by a non-equilibrium system; called "negentropy" by Schrödinger.

Gyroscope: Device reacting to changes of direction. The simplest example is a mechanical spinning top.

Heisenberg uncertainty relation: In classical mechanics, momentum ($p = mv$) and position (x) are ordinary variables that commute: $xp = px$, i.e., $xp - px = 0$. In quantum mechanics, this is no longer true. Position x and momentum p are operators that do not commute: $xp - px = i\,h/2\pi$. Since the Planck constant is very small, its importance only shows up at the atomic level. This implies, e.g., that *the order* of measuring position and velocity matters at the quantum level.

Interference: Interaction of waves; can be constructive or destructive.

Ionic bond: See Bonding (molecules).

Lattice: Regular arrangement of nodes, e.g., atoms.

Lattice vibration: Oscillations of nodes (atoms) in a lattice.

Lorentz transformation: Describing the mass of a particle as a function of its velocity, taking into account the speed of light. The basis for Einstein's theory of relativity.

Lysosome: Vesicle transporting neurotransmitters and other subjects.

Maxwell's equation: Describes mathematically the dynamics of electric and magnetic fields in terms of the motion of electric charges.

Micrometer (µm): 10^{-6} (0.000 001) meters—a millionth of a meter. (See Table of Units). Biological cells are typically 10 µm across. The wavelength of the green light in a laser pointer is 0.532 µm.

Microsecond (µs): 10^{-6} (0.000 001) seconds—a millionth of a second. (See Table of Units).

Microwaves: Electromagnetic waves with frequencies in the GHz regime, used e. g. in microwave ovens and in wireless connections in the home. Their wavelength is on the centimeter scale.

Millimeter (mm): 10^{-3} (0.001) meters—a thousandth of a meter. (See Table of Units).

Millisecond (ms): 10^{-3} (0.001) seconds—a thousandth of a second. (See Table of Units).

Molecule: Aggregate of atoms. From small simple molecules like for example, oxygen—two oxygen molecules with a covalent bond (O_2, O=O)—to large proteins and hormones made up of thousands of atoms.

Nanometer (nm): 10^{-9} (0.000 000 001) meter—a thousandth of a micrometer. (See Table of Units). Atoms are typically 0.1–0.3 nm in diameter. The membrane of biological cells is typically 6 nm thick.

Nanosecond (ns): 10^{-9} (0.000 000 001) seconds—a thousandth of a microsecond. (See Table of Units).

Nanotechnology: Technology where natural or manufactured structures with sizes from 1 to 100 nanometers are handled. The transistors in a modern microprocessor are 5–10 nanometers in size.

Neutron: Neutral (uncharged) elementary particle. Along with protons one of the constituents of the atomic nucleus.

Newton's equation: Describes the acceleration a of a particle with mass m under the influence of a force F:

$$F = m\,a = m\frac{dv}{dt} = m\frac{d^2x}{dt^2}.$$

Nucleotide: Molecules built from amino acids and phosphate molecules. Forms the rungs of the DNA ladder and represents the letters A, T, C, and G of the genetic code.

Oscillation: Motion characterized by amplitude, frequency, wavelength, propagation rate, phase, coherence, and decoherence.

Pauli principle: Two electrons cannot be in the same state at the same place at the same time. Specifically, two electrons occupying the ground state orbital of the hydrogen molecule must have opposite spin.

Particle-wave duality: The quantum physical phenomenon that particles can behave as waves, and waves as particles.

Periodic table: Map where the elements are sorted according to the electron structure of the respective atoms and thus their chemical properties.

Phonon: Single quantum of vibration.

Photoelectric effect: Means that the energy of a photon is sufficient to ionize an atom by detaching an electron. This is the explanation that a photocell / solar cell can absorb light and drive an electric current.

Photon: Quantum of EM radiation.

Picometer (pm): 10^{-12} (0.000 000 000 001) meters—a thousandth of a nanometer. (See Table of Units).

Picosecond (ps): 10^{-12} (0.000 000 000 001) seconds—a thousandth of a nanosecond. (See Table of Units).

Protein: Large organic molecule made up of amino acid chains. Thousands of different proteins make up most living things.

Proton: Elementary particle with positive charge +1. Builds up the atomic nucleus together with neutrons.

Quantum bit: Quantum mechanical system with two possible levels. Most easily represented by the two levels of an electron in a magnetic field ("spin up / spin down").

Quantum computer: A computer where the memory is made up of quantum bits that can be put in superposition and entangled via external control systems, such as microwave pulses. Calculations are performed directly in the memory and controlled by algorithms created in a standard classical computer.

Quantum physics: The doctrine of physics at the atomic and molecular level where electrons behave like waves, and where electromagnetic fields must be described in terms of photons.

Quantum tunneling: Passage of a quantum particle (e.g., an electron) through a barrier.

Quantum mechanics: Mathematical description of quantum physics.

Resonance: Self-oscillation in a system such as a piano string or an organ pipe.

Schrödinger equation: Quantum version of Newton's equation.

$i\,(h/2\pi)\,d\psi/dt = -(h/2\pi)^2/2m\,d^2\psi/dx^2 + V(x)\,\psi.$

Schrödinger's cat: The victim in a thought experiment where the cat could formally be put in a super position state of being alive and dead. The thought experiment was that the idea was absurd.

Signal substance: Molecules that are transported between different parts of a cell or between cells, such as brain nerve endings (synapses), and that initiate biochemical reactions.

Spectroscopy: The study of a wave motion spectrum of frequencies that can be used, e.g., for the analysis of chemical substances.

Speed: The size of the velocity, not caring about its direction. Speed is a scalar—the length of the velocity vector.

Spin: The property of elementary particles, for example electrons, which have no direct equivalent in classical physics. The term spin sounds like a kind of rotation, but it is rather a magnetic property that can have a number of definite, quantified, values. An electron has two such values: +1/2 and –1/2. In a magnetic field, the electron can be in a superposition of two states.

Superposition: Refers to addition of quantities. The wave function of a quantum particle is often described as a superposition of quantum mechanical states—the particle-wave is simultaneously in several states. Like hitting several keys on a piano to create a chord. When playing on quantum physical instruments, one creates superpositions of quantum states. These can be different electron states in an atom, different polarizations of a photon, or different directions of an electron's spin in a magnetic field.

Tunneling: Wave passage through a barrier. For instance, passage of light between two optical fibers in contact with each other.

Turing Test: Alan Turing 1950 reformulated "can a machine think" in terms of a kind of intelligence test: "can a digital computer imitate a human being."

Velocity: Velocity describes the speed and the direction of an object. The velocity is a vector.

Vesicle: Liquid-filled spherical container limited by a cell membrane, an "empty cell." A vesicle is often formed from an indentation of a cell wall. Works as a transport vehicle in or between cells. The lysosome is a vesicle.

X-ray radiation: Electromagnetic radiation with high energy and wavelengths in the range of 0.01–10 nm. The boundaries are blurred: soft X-rays border on ultraviolet radiation, and very hard X-rays border on gamma radiation.

Table of Units

Short scale—used in most English and Arabic-speaking countries. The short scale is used in this book.

Long scale—used in continental Europe and most French-, German-, and Spanish-speaking countries.

| Prefix | | Base 10 | Decimal | English word | | Adoption[nb 1] |
Name	Symbol			Short scale	Long scale	
yotta	Y	10^{24}	1 000 000 000 000 000 000 000 000	septillion	quadrillion	1991
zetta	Z	10^{21}	1 000 000 000 000 000 000 000	sextillion	trilliard	1991
exa	E	10^{18}	1 000 000 000 000 000 000	quintillion	trillion	1975
peta	P	10^{15}	1 000 000 000 000 000	quadrillion	billiard	1975
tera	T	10^{12}	1 000 000 000 000	trillion	billion	1960
giga	G	10^{9}	1 000 000 000	billion	milliard	1960
mega	M	10^{6}	1 000 000	million		1873
kilo	k	10^{3}	1 000	thousand		1795
hecto	h	10^{2}	100	hundred		1795
deca	da	10^{1}	10	ten		1795
		10^{0}	1	one		–
deci	d	10^{-1}	0.1	tenth		1795
centi	c	10^{-2}	0.01	hundredth		1795
milli	m	10^{-3}	0.001	thousandth		1795
micro	μ	10^{-6}	0.000 001	millionth		1873
nano	n	10^{-9}	0.000 000 001	billionth	milliardth	1960
pico	p	10^{-12}	0.000 000 000 001	trillionth	billionth	1960
femto	f	10^{-15}	0.000 000 000 000 001	quadrillionth	billiardth	1964
atto	a	10^{-18}	0.000 000 000 000 000 001	quintillionth	trillionth	1964
zepto	z	10^{-21}	0.000 000 000 000 000 000 001	sextillionth	trilliardth	1991
yocto	y	10^{-24}	0.000 000 000 000 000 000 000 001	septillionth	quadrillionth	1991

Index